AF692315

Telecommunications Technologies for Small Business

Frank Panzarino

ISBN 0-13-093742-8

90000

9 780130 937420

PH PTR

Prentice Hall PTR
Upper Saddle River, NJ 07458
www.phptr.com

Library of Congress Cataloging-in-Publication Data

Panzarino, Frank
Telecommunications technologies for small business / Frank Panzarino.
p. cm.
Includes index.
ISBN 0 13-093742-8
1. Telecommunication. 2. Telecommunication—Equipment and supplies. 3. Small business—Data processing. I. Title.

TK5101 .P313 2001
621.382—dc21 2001036969

Cover design director: *Jerry Votta*
Cover design: *Design Source*
Manufacturing buyer: *Maura Zaldivar*
Acquisitions editor: *Victoria Jones*
Editorial assistant: *Michelle Vincenti*
Marketing manager: *Debby van Dijk*

Prentice-Hall, Inc.
Upper Saddle River, NJ 07458

Prentice Hall books are widely used by corporations and government agencies for training, marketing, and resale. The publisher offers discounts on this book when ordered in bulk quantities. For more information, contact Corporate Sales Department, Phone: 800-382-3419; FAX: 201-236-7141;
E-mail (Internet): corpsales@prenhall.com
Or write: Prentice Hall PTR, Corporate Sales Dept., One Lake Street, Upper Saddle River, NJ 07458.

Printed in the United States of America
10 9 8 7 6 5 4 3 2 1

ISBN 0-13-093742-8

Pearson Education, LTD.
Pearson Education Australia PTY, Limited
Pearson Education Singapore, Pte. Ltd.
Pearson Education North Asia Ltd.
Pearson Education Canada, Ltd.
Pearson Educación de Mexico, S.A. de C.V.
Pearson Education—Japan
Pearson Education Malaysia, Pte. Ltd.

To my Mom, who showed me by example that hard work produces results.

Contents

Preface

Surviving the rigors of small business requires knowledge in personal computing and telecommunications. *Telecommunications Technologies for Small Business* provides you with the basic tools to survive and prosper in a small business or home environment.

Many of today's technologies are evolving at a rapid pace. It is difficult (if not impossible) to remain current in the growth areas of high-speed broadband networking. This book guides the reader to a general understanding of how networks and devices are interconnected and what the latest trends in the industry are. Telecommunications equipment, including the personal computer, is examined for the purposes of benefiting both the home and small business user.

Telecommunications Technologies and the Structure of This Book

In Part 1 – Telecommunications Systems and Services, we start simply, by reviewing the basics of electronics theory and examining the existing telephone network: its outstanding benefits and its limitations. Analog and digital telecommunications networks are emphasized along with standard telephone network technologies. Audio and video services, including both cable and satellite transmission systems, as well as high-speed broadband Internet access are topics also introduced in this Part.

Part 2 – Telecommunications Voice and Data Terminal Equipment provides knowledge important for personal and business computing. In this Part, the equipment and services needed for a standard office installation are described. Software needed to create a home office or small business environment is also reviewed. Security considerations are an often overlooked but important part of the necessary requirements for small business installations. Unfortunately, the disaster at the World Trade Center in New York City illustrates the need for backup and recovery systems and networks. Privacy and security implications are covered in this Part. Other devices, such as answering machines, cordless telephones, and telephone features important to home or small business environments, are also reviewed here.

The last Part of this book takes a look at the new and upcoming telecommunications equipment and services. Part 3 – Telecommunications Technologies for the Future provides a view into the new technologies that will be available for the telecommunications environment of the future. Faster speeds, faster switching, and home integration are some of the topics covered in this Part. Wireless technologies will certainly be an important area of future telecommunications services. Computer interfaces for the home environment are in development at this time. Integration and convergence of existing technologies will have a profound impact on our lives in the future.

Throughout the book, I kept formulas and exhaustive analytical analysis to a minimum, to keep the book readable, informational, and at the same time enjoyable.

Cellular services are not covered in any detail, since the intention of the book is to provide a basic understanding of telecommunications technologies for small business and not for a constantly mobile business environment. Also, some high-end technologies such as ATM and SONET are not reviewed in this book since these technologies are used for telecommunications networks or large business applications and would not be encountered in a standard home or small business environment.

Telecommunications Technologies for Today's Environment

Telecommunications technologies and services are synonymous with change. We are accustomed to more and more information at a lower cost. We desire the latest information available anytime, anywhere in any environment.

Because of these changes in technology, human communication has also changed in a short span of time. The social, business, and economic environments are evolving with the technology, and this impacts us as well. For example, email has replaced written letters; long distance voice has also replaced many forms of written communication. We can now video-conference, instant-message, and exchange pictures and music with relative ease. The integration of this technology into our society has had a profound impact on our lives. We must embrace today's technology in order to remain competitive.

Telecommunications Technologies for Classroom Instruction

Telecommunications Technologies for Small Business can also be used as an introduction to today's telecommunications services and technologies for both students and business professionals. High schools, technical schools, and colleges can benefit from this book for their introductory technology and networking courses.

I have included many illustrations to emphasize topics that will benefit students. The tables and images combined with the text are a good source of information for someone with a desire to learn and study. The book begins with the basics and moves on up. A teacher or professor can emphasize certain sections of the book while deemphasizing others. For example, students can focus on multimedia audio and video transmission.

I've taught introductory telecommunications courses at several schools and received feedback from many students. Students may have knowledge in one particular area of telecommunications and networking technology but not a complete introductory knowledge of voice, video, and data. It is helpful for students to see the "Big Picture" before diving into the bits and bytes of data transfer.

This book is ideal for nonengineering or computer science majors who need to know the essentials of telecommunications technologies.

Telecommunications Technologies—Acknowledgments and Thanks

It's been a great experience writing this book. You learn by writing down your thoughts and ideas, and this book is no exception. I've researched several textbooks, articles, web sites, documents, and news clippings and drawn from my personal experiences in the telecommunications industry to complete this book.

I'd like to acknowledge the Wesley J. Howe School of Technology Management at Stevens Institute of Technology in Hoboken, New Jersey. My association with the Telecommunications Management program at the college goes back over 10 years, first as a student and then as an adjunct professor. Specifically, I'd like to thank Dr. Kazem Sohraby and Dr. Kevin Ryan, who have encouraged me and provided me with the opportunity to teach and develop introductory telecommunications courses for the college. It continues to be a great experience.

I'd also like to acknowledge my friends and coworkers at MCI Worldcom, Inc. Some of us have shared many experiences together, both technical and otherwise. We continue to survive and thrive in this fast-paced business we know as telecommunications. Thanks for the knowledge and great memories.

Many, many thank-you's to my family:

My father, who instilled in me an enduring work ethic.

My daughter Vanessa, who compliments me as a father when she wants her college major to be engineering.

My step-son Christian, who impresses me with his goals in police and detective work.

My wife Beth, who put up with months of waiting for me to finish writing this book. You constantly give me strength when I begin to falter.

Telecommunications Technologies and Reader Feedback

I'll always welcome constructive feedback from readers. You can contact me directly through personal email:

fpanzari@engineer.com

or through Prentice Hall PTR:

ptr_feedback@prenhall.com

I'll do my best to respond promptly. Read on and enjoy!

Frank Panzarino

About the Author

Frank Panzarino's telecommunications experience spans more than 20 years: first with RCA Global Communications Inc., and more recently with MCI Worldcom Inc. He has held various positions within these companies including technician, programmer, project manager, and engineer. Mr. Panzarino also began a teaching career at DeVry Institute and is currently an adjunct professor at Stevens Institute of Technology in Hoboken, New Jersey. His teaching duties also include the planning, preparation, and creation of introductory courses for the Telecommunications Management program at the college.

Part 1

Telecommunications Systems and Services

1 Telecommunications Basics

The basic purpose of any telecommunications system is to convey a message from one point to another. Electronic circuits and wired networks are used to transmit and receive information to attain this goal. Electronic circuits also play a role in storing, encoding, decoding, and processing information.

With advances in electronics technology, user equipment rapidly becomes obsolete and networks become faster, cheaper, and smarter. Advancements in microprocessors continue to reduce the size and sophistication of these devices needed to transmit and receive information.

The computer industry has also spawned another industrial revolution, and we have now moved from the electronics age to the information age. Computers have uses ranging from simple everyday tasks, such as calculating bills, to more complex scientific tasks, such as analyzing data. In telecommunications, computers are used as end-terminal devices needed to communicate in today's business environment.

However, computers alone cannot tell the story because we must communicate outside our local environment. This is why telecommunications is so important for today's businesses. Modems, routers, and switching terminals enable us to extend our information outside our user end-terminal device (computer).

Before we discuss the latest trends and advancements in telecommunications and technologies for small businesses, we must understand the basics needed to create simple telecommunications circuits. We will also closely examine the wiring needed to transmit information from one point to another.

This chapter is divided into five sections. The first section covers basic electronics circuit theory including resistance, current, voltage, and power. The second section covers frequency, modulation, and bandwidth, with explanations and examples of each. The third section looks briefly at noise, attenuation, and crosstalk in a circuit. The fourth section covers guided transmission media; specifically, twisted pair, coaxial cable, and fiber optic cable. The fifth and final section covers unguided transmission, which includes the airwaves.

Here we go!

BASIC ELECTRONICS CIRCUIT THEORY ....

The concepts of basic electronics circuit theory apply to telecommunications circuits and systems. While today's networks are very sophisticated with intelligent equipment and fiber lines, a circuit containing resistance, voltage, and current is always necessary. Let's start by examining the basic copper wire.

Conductors, Nonconductors, and Semiconductors

The basic copper wire is a conductor of electricity. A conductor contains material that can transfer an electrical charge from one object to another. Examples of conductors are copper and aluminum. Wire always contains a conductive material inner core, enabling the flow of electrons. An insulator (or nonconductor) contains material that opposes the flow of charge. Wire is almost always covered with an insulator. Materials such as plastic, wood, and ceramics are examples of good insulators. The wire cover or shroud contains insulating material so that wires cannot touch each other.

Materials such as semiconductors contain the elements of both conductors and semiconductors. An example of an item with this type of material is a transistor. A transistor is another name for a semiconductor.

Now that we have determined that an electrical charge can move through a conductive material, we must determine other electrical requirements to build a circuit.

Current in a Circuit

Electric current is defined as the flow of charged particles through a copper wire or a conductor in a specified direction. Charged particles flowing in a copper wire are called electrons. Electron flow is from the negative terminal of a power supply to the positive terminal of a power supply. A good analogy of electron flow is the water flow in a pipe inside your residence. Think of the water as electrons and the pipe as a copper wire.

Is the water flowing rapidly or is the flow weak or nonexistent? The same questions apply to an electrical circuit. If the electrical circuit contains a battery, the current may be flowing rapidly in the beginning but as the battery weakens, the current flow may be weak. Of course, a dead battery will produce no current flow at all.

There are two types of current: Direct Current (DC) and Alternating Current (AC). Direct current has the same direction or flow. A battery can be considered a direct current source. This flow is from negative to positive, as previously described. Direct current can also be described as unidirectional.

Alternating current varies with time. Current in a home or apartment is AC, operating at 60 cycles per second. Each complete cycle of current flow contains current flowing in the positive range for one-half of the cycle and in the negative range for the other half of the cycle. This positive-to-negative cycling is the equivalent of a sine wave occurring 60 times in one second.

The unit of measurement for current is amperes, or amps (A).

Voltage in a Circuit

The voltage of a circuit is the difference in potential between any two charges that are exactly equal to each other. Voltage can be described as the circuit pressure. A good analogy is the water pressure of a pipe. The flow of water is equally important as the pressure.

A voltage source, such as a battery, converts one form of energy into electrical potential energy. This voltage source also contains a positive and a negative charge. A battery can be considered a voltage source that contains a stored chemical energy. Batteries provide conversion from chemical energy into electric energy, which provides voltage and current for a circuit. In high voltage systems, the negative terminal of a voltage source is usually grounded to earth.

The unit of measurement for voltage is volts (V).

Resistance in a Circuit

Resistance is the amount of opposition to current flow. A wire that is not a good conductor will provide resistance to current flow and, therefore, will affect the overall performance of a circuit. The resistance of wire is determined by the following:

- Type of material used as a conductor (Table 1.1)
- The length of the material
- The cross-sectional area of the material (Table 1.2)
- The temperature of the material

The higher the temperature of a conductor, the less resistance encountered. When pure metals are used, then resistance is generally proportional to the temperature. For example, copper at 100°C has approximately 43% less resistance than at 0°C.

Wires have sizes called the American Wire Gauge (AWG). This gauge sets the standards for the cross-sectional area of any material. In general, an increase of three gauge numbers multiplies the resistance by a factor of two. Conversely, a decrease of three gauge numbers doubles the area and decreases the resistance by half. It is always desirable to use the smallest (and least costly) wire that a circuit can use. If the wire is too small, the current will not be able to flow because of the resistance of the wire and the wire may open or burn up.

A wire's length will contribute to a loss in signal. This is referred to as weakening of a signal or attenuation of a signal. Attenuation is an important factor in telecommunications circuit design. In general, the resistance of a conductor or wire is directly proportional to its length and inversely proportional to its cross-sectional area.

The unit of measurement for resistance is ohms (Ω).

Table 1.1 Resistance of Materials

Material	Resistance	Conductivity
Aluminum	$2.83 \times 10^{-8}\,\Omega$	Conductor
Copper	$1.72 \times 10^{-8}\,\Omega$	Conductor
Germanium	$47 \times 10^{-2}\,\Omega$	Semiconductor
Gold	$2.45 \times 10^{-8}\,\Omega$	Conductor
Mica	$2.02 \times 10^{-10}\,\Omega$	Insulator
Silicon	$64 \times 10^{2}\,\Omega$	Semiconductor
Silver	$1.64 \times 10^{-8}\,\Omega$	Conductor
Teflon	$3 \times 10^{15}\,\Omega$	Insulator

An electrical load can be a device connected across an energy source. Loading devices are resistive and oppose current flow. For example, a telephone terminal can be considered a loading device. This device, including its associated internal components, can be considered a resistive load in a circuit.

Table 1.2 lists the AWG and resistance for copper.

Table 1.2 American Wire Gauge (AWG) and Resistance for Copper

AWG	Diameter (Mil)	Area (Cm)	Resistance (Ω/kft)
0	324.9	105,330	0.0983
1	289.3	83,694	0.1240
2	257.6	66,373	0.1563
3	229.4	52,634	0.1240
4	204.3	41,742	0.1563
5	181.9	33,102	0.313
6	162.0	26,250	0.395
7	144.3	20,816	0.498
8	128.5	16,509	0.628
9	114.4	13,094	0.792
10	101.9	10,831	0.998
11	90.74	8,234	1.260
12	80.81	6,529.9	1.588
13	71.96	5,178.4	2.003
14	64.08	4,106.8	2.525
15	57.07	3,256.7	3.184
16	50.82	2,582.9	4.016
17	45.26	2,048.2	5.064
18	40.30	1,624.3	6.385
19	35.89	1,288.1	8.051
20	31.96	1,021.5	10.15

Ohm's Law

Ohm's Law is a formula that can be used to find the current, voltage, or resistance of a circuit. Ohm's Law states that when the voltage in a circuit increases, the current also increases in direct proportion to the applied voltage. Therefore, a constant ratio between voltage and current exists.

$$\text{Ohm's Law} = \text{I (current)} = \text{E (voltage)} \div \text{R (resistance)}$$

or

$$I = E \div R$$

By manipulation of Ohm's Law, any one of the three variables can be determined if the other two are known. For example, if the voltage of a circuit is 12 volts and the resistance of the same circuit is 10 Ω, the current must be 1.2 amperes (or 12 divided by 10 equals 1.2), as seen in Figure 1.1.

Power in a Circuit

Power is the rate at which work is done or energy expended. It is an indication of how much work can be accomplished in a certain amount of time. In a circuit, the current and voltage are multiplied together. The resulting product is the amount of power dissipated in a circuit. The unit of measurement for power is the watt.

$$\text{Power} = \text{P (watts)} = \text{E (voltage)} \cdot \text{I (current)}$$

For example, if the voltage of a circuit is 12 volts and the current of the same circuit is 1.2 amperes, the power must be 14.4 watts (or 12 multiplied by 1.2 = 14.4). Notice how Ohm's Law is manipulated again to obtain the result for power (Figure 1.1).

Decibels

The decibel (dB) was defined by the telephone company to express the gain or loss of a signal in telecommunications networks. When the output power and the input power are equal, the power ratio is 1 and the decibel ratio is 0 dB. A decibel uses a logarithmic scale, not a linear scale. For example, a doubling of power would increase the circuit by 3 dB. Similarly, a loss or decrease of power (by half) would decrease the circuit by –3 dB.

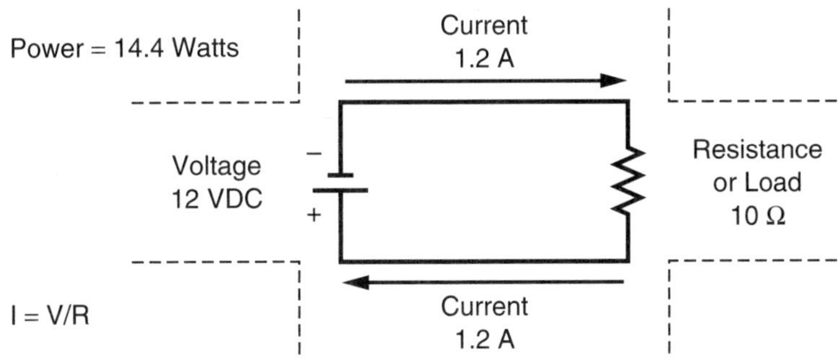

Figure 1.1
Power and Ohm's Law in a circuit.

Because decibels are logarithmic, the gains expressed in dB can be added (or subtracted) instead of multiplied (or divided). Here is an example:

An antenna amplifier has a gain of 3 dB and the coaxial cable connected to the antenna has a loss of 1.5 dB. The cable is connected to a tuner with a gain of 36 dB, which is then connected to a 9 dB amplifier. The speakers provide a loss of 6 dB. Simple addition/subtraction would produce a $3 - 1.5 + 36 + 9 - 6 = 40.5$ dB gain.

Noise in a Copper Wire

Noise is always present in copper wire. Long runs of copper wire behave as an antenna. An antenna can receive spurious unwanted signals and can also transmit spurious unwanted signals (under the right conditions). For example, a truck that isn't properly tuned can emit signals that can be received by copper telephone wires. These types of signals, called impulse noise, can sometimes be heard as pops or clicks in an analog telecommunications circuit.

Telecommunications equipment, power supplies, fluorescent lighting, and AC wiring are other examples of noise that can be injected into a circuit. Proper design of the equipment and the network itself keeps noise levels to a minimum.

Signal-to-Noise Ratio

Let's take the concept of a decibel one step further. We can compare an analog signal relative to the background noise and perform a ratio of the two signals. This type of signal measurement is called the signal-to-noise ratio and is commonly used in telecommunications circuits.

A good analogy of the signal-to-noise ratio can be made with reference to audio equipment. An analog record may have a 50-dB signal-to-noise ratio but a digital compact disk (CD) may have a 95-dB signal-to-noise ratio. Our ears can discern a difference in the audio quality between a record and a compact disk. The overall quality of a digital signal is inherently better also. We'll discover why this is so in Chapter 3.

A Basic Telecommunications Circuit

Now that we have an understanding of the basic components that comprise an electronics circuit, let's apply this knowledge to a practical telecommunications circuit (Figure 1.2).

The telephone company switching center contains local telephone lines and equipment that is needed for standard telephone service. This center contains a 12-volt battery source needed for today's voice communications. These batteries also have a backup generator, which can generate 12 volts in the event of a blackout or power failure. Note: Your telephone service will usually function even when the lights in your home or apartment do not.

The 2-wire local loop provides access from the telephone company switching center to your residence. This wire is usually 24-gauge copper and is contained in large bundles on a telephone pole or under a street. The job of the local telephone company is to provide you with a working pair of these wires to complete the circuit.

Your residence has a telephone set that acts as a resistive load on the circuit when it is in use. If you have an extension phone, try making a call and then have someone pick up the extension and listen. You will hear the loudness decrease because a second resistive load has been added. Therefore, the circuit resistance has been increased. Try picking up additional extension phones (if you have them) and note what happens.

We discuss more about the switching center, 2-wire local loops, and the telephone handset device in Chapter 2.

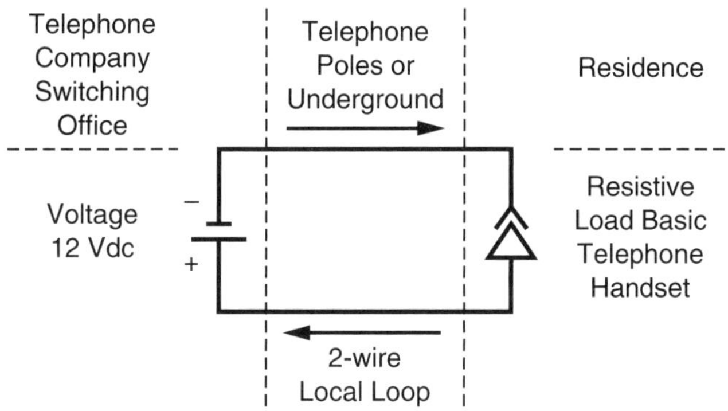

Figure 1.2
Basic telecommunications circuit.

TIMETABLE OF HISTORIC ELECTRICITY AND ELECTRONICS EVENTS

1733	Charles Du Fay discovers that a glass rod becomes electrified when rubbed with silk.
1750	Benjamin Franklin discovers electricity with his famous kite and key experiment during a thunderstorm.
1785	Charles Augustin de Coulomb proves the laws of attraction and repulsion exist with electric charges.
1786	Luigi Galvani discovers that a frog's leg could move if copper contacts a nerve or muscle.
1796	Alessandro Volta proves that electricity could be produced if unlike metals separated by moistened paper are brought into contact with each other.
1820	Hans Christian Oersted discovers that a current-carrying wire produces magnetic fields. This behavior was later known as electromagnetism.
1826	Georg Simon Ohm observes that electric resistance remains constant over wide ranges of potential difference. This observation was later known as Ohm's Law.
1860	Michael Faraday and Joseph Henry publish the laws of electromagnetic induction findings.
1868	Heinrich Geissler develops tubes in which electrical charges in gas produce different colors.
1880	Sir William Crookes invents the cathode ray tube (CRT).
1883	Thomas Edison's findings produce the first thermionic diode.
1904	Sir John Fleming develops the Fleming valve or vacuum tube rectifier.

TIMETABLE OF HISTORIC ELECTRICITY AND ELECTRONICS EVENTS (CONTINUED)

1907	Lee DeForest invents the first vacuum tube capable of amplifying small electrical signals.
1946	J. Pesper Eckert and John Mauchly create the ENIAC electronic computer by using vacuum tubes.
1947	John Bardeen, Walter Brattain, and William Shockley invent the transistor.
1958	Jean Hoerni, Jack Kilby, Kurt Lehovec, and Robert Noyce develop the first integrated circuit (IC). The IC contains thousands of transistors on a small semiconductor chip.
1971	Ted Hoff invents the microprocessor, which greatly reduces the size of computers and computing devices.
1977	Apple, Radio Shack, and Commodore market the first personal computers.
1979	Motorola produces a powerful 16-bit microprocessor.
1993	Intel produces a 64-bit Pentium microprocessor.

FREQUENCY, MODULATION, AND BANDWIDTH

The basics of telecommunications transmission require analog frequency signals that can be modulated and then transmitted. Modulation is an information frequency signal. For example, the human vocal cords are modulated to produce human speech. Human speech can be considered the information signal you want to transmit.

There are several different types of transmission techniques. Some of these have been used for more than 60 years. Let's first begin by discussing what frequency is.

Frequency

Frequency is the number of occurrences of a periodic event such as the cycle of a waveform, a vibration, or oscillation. For example, a steady alarm tone from a firehouse may be a loud frequency that we can hear to alert us of a fire.

Frequency is strictly analog. Sound waves generated by music and human speech have complex frequency characteristics that we can hear. These frequency characteristics are the number of vibrations per second that will provide a particular sound. These

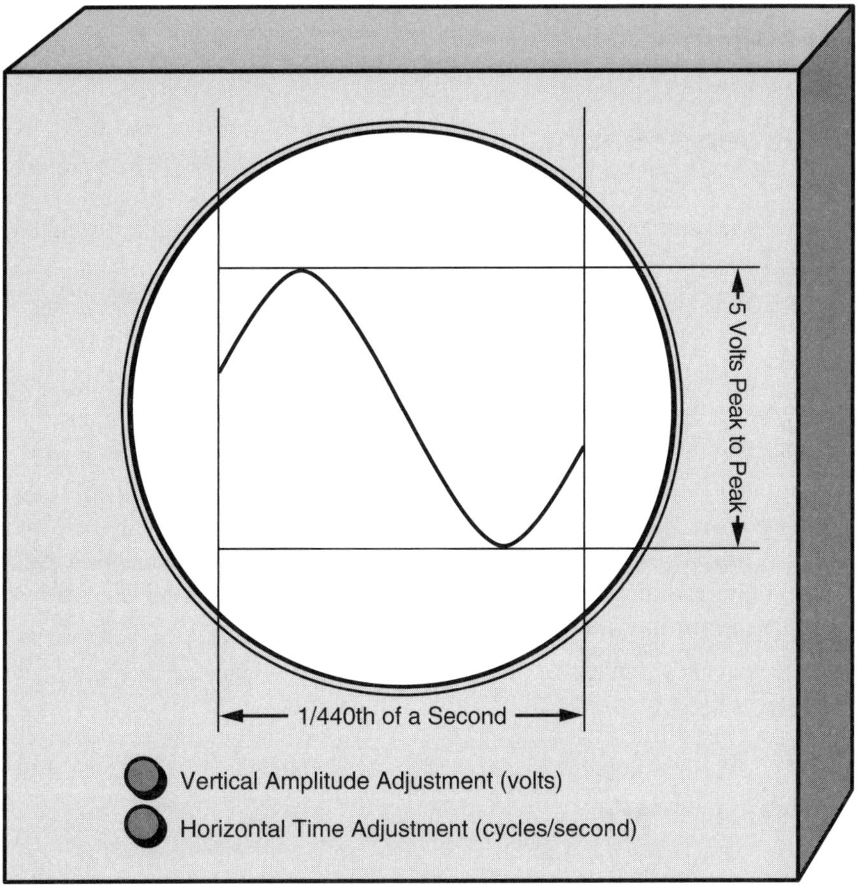

Figure 1.3
440 Hz A note observed on an oscilloscope.

frequency characteristics are identical to the frequency characteristics used for transmission over guided (wires, cables, etc.) and unguided (air, water, etc.) media.

To examine signal frequency more closely, let's view a steady tone such as the tone from the firehouse alarm. A steady tone at any frequency would appear as a sine wave on an oscilloscope. An oscilloscope is a device that can view the amplitude and frequency of a signal.

Another example of frequency would be the A note on a piano. This note would appear as 440 sine wave cycles per second on an oscilloscope (Figure 1.3). In other words, if we were able to view one second of time, we would see 440 cycles of the sine wave. The frequency of this signal is measured in Hertz (Hz for short). Hertz is the equivalent of cycles per second.

Each sine wave has a hill and a valley, which define the amplitude of the signal. As the signal gets greater (or louder), the amplitude of the sine wave gets greater.

Frequency Range

Frequencies can be expressed as the lowest signal and the highest signal that can be used. This is referred to as the frequency range.

For example, a sound-producing instrument such as the piano has the lowest note on the left side of the keyboard as its lowest frequency and the highest note on the right side of the keyboard as its highest frequency.

In other words, any sound-producing instrument has a lowest and a highest frequency that it can produce.

What about telecommunications frequency ranges? Yes, at the time of transmission we must determine the highest and lowest frequencies we can use over the media we are working with. The frequency ranges for telecommunications media such as wire, fiber, and airwaves are generally much broader than for sounds we can hear, such as instruments, human voice, etc.

Frequency ranges of telecommunications media are revisited later in this chapter.

Analog Music

The frequency range of music is from 20 Hz to 20,000 Hz (Figure 1.4). Between 15,000 Hz and 20,000 Hz, most adults can sense the sound but not actually hear it. Good high fidelity receivers can reproduce both the highest and lowest frequencies with faithfulness to the original signal recorded.

Analog Human Speech and Hearing

The frequency range of human speech is between 100 Hz and 10,000 Hz (Figure 1.4).

The frequency range of human hearing decreases with age. Generally, a child can hear a range of frequencies from 20 Hz to 20,000 Hz. An adult can hear a range of frequencies from 20 Hz to 15,000 Hz (Figure 1.4).

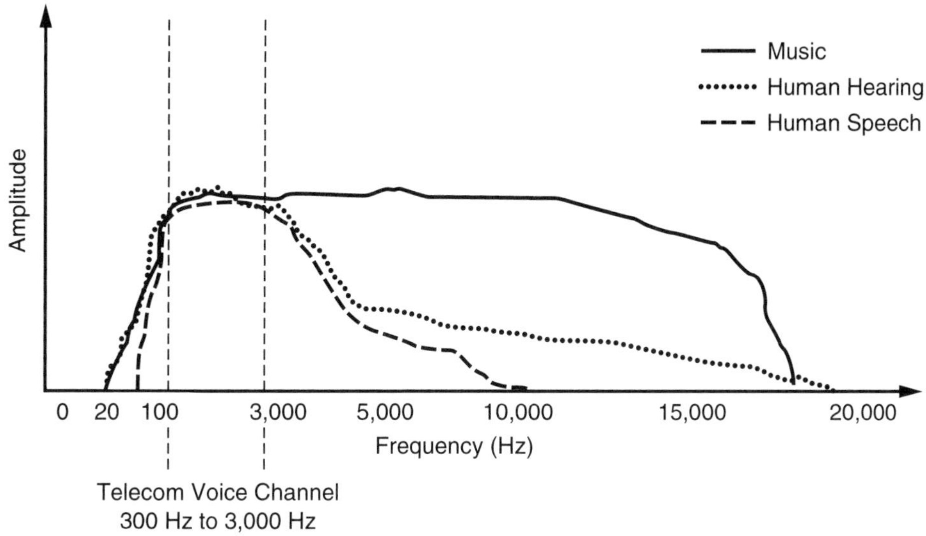

Figure 1.4
Human speech/hearing, music, telecom voice channel.

Analog Telecommunications Voice Transmission

The frequency range of an analog telecommunications voice circuit is much narrower than that of any of the ranges described above. Why? Well, the early developers of telecommunications voice networks had to decide how much voice information they wanted to transmit. Human speech has a frequency range between 100 Hz and 10,000 Hz. If all of the voice information were to be transmitted, a wide frequency range would be allocated for each telephone transmission. This would prove to be too costly and inefficient. Therefore, a narrower frequency range would be chosen on the basis of sound quality and intelligibility.

Basically, the frequency range chosen was the center frequencies of the human voice spectrum. 300 Hz to 3,000 Hz provided the voice clarity needed for telephone service yet kept the frequency range at a minimum.

This is an important fact. Analog voice transmission systems still have a frequency range limitation of 300 Hz to 3,000 Hz and therefore do not transmit the entire human voice frequency spectrum (Figure 1.4).

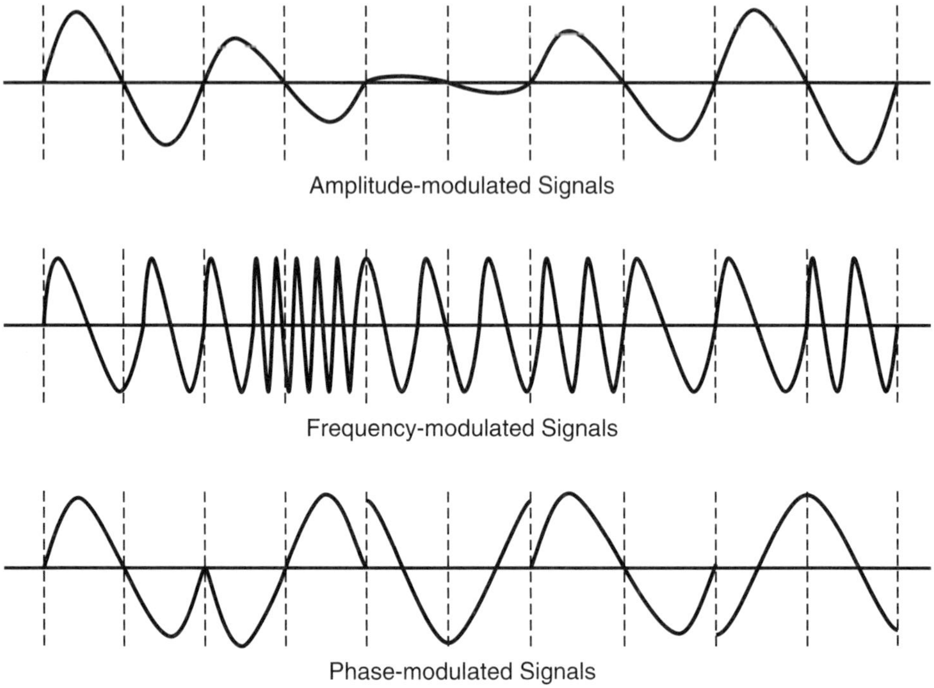

Figure 1.5
Modulation.

Modulation

Modulation is defined as a change or stress in vocal chords. In music, modulation may be a change to a different key on the music scale. In telecommunications, modulation is defined as a change of one or more parameters of a carrier wave (Figure 1.5).

Modulation of a carrier wave is used in both guided and unguided transmission systems. Each modulated signal contains a carrier wave, which provides a source frequency. Amplitude-modulated (AM) signals contain shifting amplitudes on a carrier wave in accordance with a varying modulated signal. Frequency-modulated (FM) signals contain shifting frequencies on a carrier wave in accordance with a varying modulated signal.

Shifting the amplitude (as in AM) or shifting the frequency (as in FM) are important examples of modulation. At the receiver, a detector must be able to remove or unscramble the modulated signal and relocate it back to the original frequencies. This process is called demodulation.

A third example of modulation is called phase shift. A signal's phase is the relative position of the sine wave measured in degrees. Normally, a sine wave starts at 0° or in-phase. For example, by omission of a portion of the wave, one-fourth of the wave has been deleted. This is an example of a 90° phase shift. Phase-shift modulation is very useful when digital information is transmitted. We discuss more about this topic in Chapter 3.

Bandwidth

Bandwidth is defined as the total frequency capacity available. It is measured in Hz (per second for digital). You can obtain bandwidth by subtracting the lowest frequency from the highest frequency. For example, analog human speech has a frequency range of 100 Hz to 10 kHz. Therefore, the bandwidth of analog human speech is approximately 10 kHz.

The diagram of bandwidth and frequency spectrum is shown in Figure 1.6. Note that you cannot exceed the bandwidth of any type of media without losing information either in the form of dropped bits (digital) or lost frequency response (analog).

Another aspect of analog frequency transmission is the baud rate. In analog terms, the baud rate is a unit of signaling speed. The speed in baud is the number of changes (from positive to negative, or vice versa) or events per second. An example of a baud rate is 300 baud or 300 changes per second. Baud rates are seldom used anymore. Digital bits per second (bps) has replaced the baud rate as an indicator of transmission speed or bandwidth.

Frequency Division Multiplexing

The bandwidth of a single transmission path can be divided into smaller blocks or channels. A good analogy is the slices of a cake. The cake would represent the entire bandwidth, and the slices of cake would be the individual channels. The channels cannot overlap because crosstalk interference will result. This type of frequency slicing is called frequency division multiplexing (FDM).

Let's say we have a bandwidth of 48 kHz to use and we need to transmit 12 separate voice calls. Each of those voice calls has a bandwidth of 4 kHz. OK, so $12 \times 4 = 48$. Easy! We have just enough bandwidth for these 12 channels. Channel #1 would have a frequency assignment of 1 to 4 kHz, and channel #2 would have a frequency assignment of 4 to 8 kHz. Channel #3 from 8 to 12 kHz. The voice frequencies or modulated data would ride on each of the channels and be demultiplexed down to their original form on the receiving end.

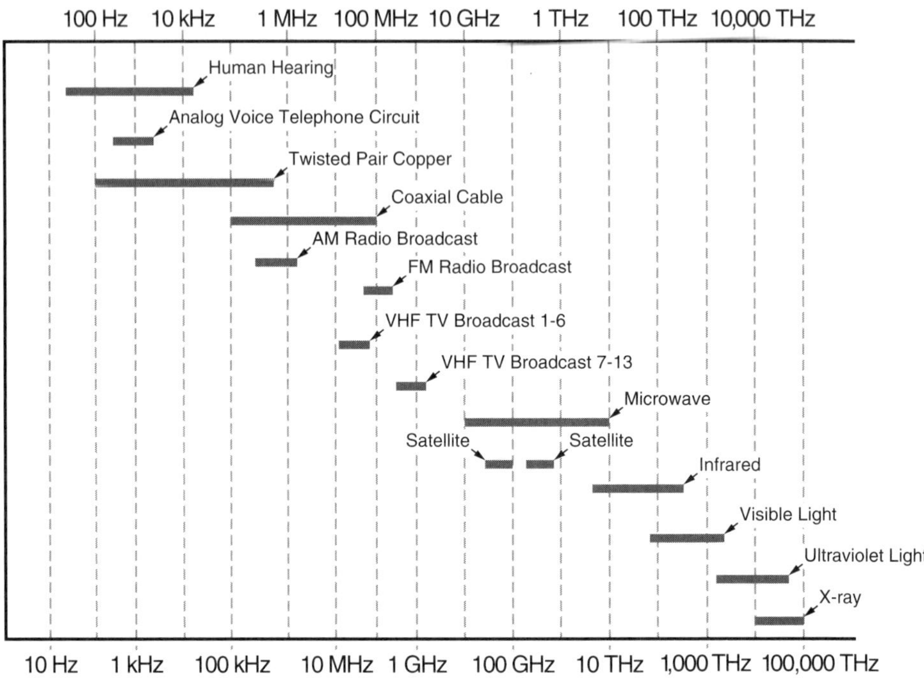

Figure 1.6
Frequency spectrum and bandwidth.

FDM has been virtually replaced by time division multiplexing (TDM), which we examine in Chapter 3.

NOISE, ATTENUATION, AND CROSSTALK..............

In an ideal situation, the transmitted signal would be duplicated over the transmission path and received correctly at the terminating end site. In reality, because of the conditions of the copper wire, equipment problems, signal paths conditions, airwave interference, weather conditions, etc., we cannot be guaranteed a reliable transmission path.

Remember, a wire is like an antenna. If the wire is long enough, it will radiate a signal (under the right conditions). A wire will also receive an unwanted signal. Noise, attenuation, and crosstalk are just some of the problems we encounter with copper wiring.

Noise

Noise is a major problem in transmission systems today. Electrical interference from wiring in a home or small office can alter the transmission signal, producing errors and signal dropouts. The solution is to keep telecommunications wiring physically separate from any electrical wiring.

Other types of noise such as impulse noise (voltage spike appearing as a click or a pop) and random noise (frequency noise from other transmission systems) are also present and must be dealt with individually.

Attenuation

Attenuation or resistance is affected by the length of wire and the wire gauge. The average copper wire gauge used in telecommunications networks is 22 AWG. There isn't much a network installer can do about attenuation, except be aware that it exists and can have an adverse effect on transmission networks.

Generally, a repeater or amplifier is used to boost a signal to acceptable levels. A repeater should be installed periodically on the wire to keep the signal levels at acceptable levels.

Crosstalk

Crosstalk refers to the coupling between adjacent wiring pairs. If telecommunications wiring is bundled together, crosstalk can result. The wires are not physically touching and some of the signal is radiated from one wire to another.

To keep crosstalk to a minimum, it is important to keep transmission dB signal levels within specifications. A signal that is running "hot" could radiate out to another wire or transmission path.

GUIDED TRANSMISSION MEDIA

We've already discussed how each type of physical transmission media (fiber, cable, and twisted pair) has a bandwidth limitation. Now, let's discuss each of the three main types of guided media (Figure 1.7): twisted pair, coaxial cable, and fiber optic. Each

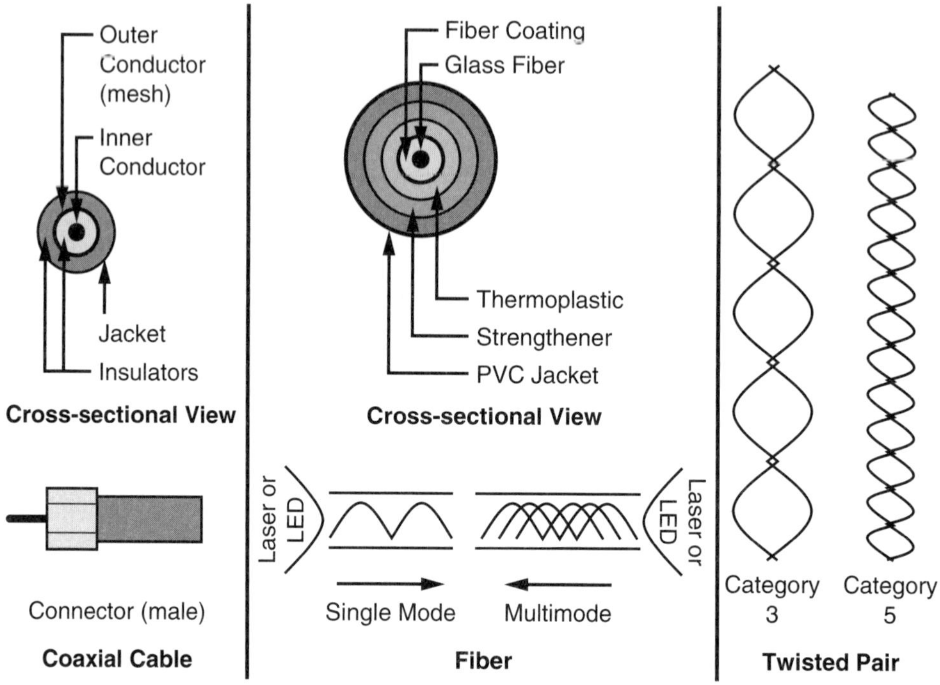

Figure 1.7
Guided transmission media.

of these transmission systems has its limitations and advantages. You must consider cost, flexibility, ease of installation, and bandwidth before using them.

Twisted Pair

Unshielded twisted pair (UTP) contains copper wires with a twist to reduce noise. The more twists per foot, the less susceptible the media will be to noise and, therefore, the greater the bandwidth. In short lengths, twisted media can support 100 Mbps or more (Figure 1.7).

The big advantage of unshielded twisted pair is its flexibility. The wires can be routed through cable trays, through conduit, and in very tight spaces in older buildings. Another advantage of unshielded twisted pair is its relatively low cost (pennies a foot).

There are different specifications of UTP. Category of performance (CAT) specification measures signals that the copper wire must pass. CAT3 can easily handle voice but may be inadequate for data. CAT3 is rated for a maximum of 16 MHz bandwidth for data. CAT5 and CAT5 extended can handle 100 MHz bandwidth for data. The main difference between CAT5 and CAT5e is that the specification has been made stricter in the new version. CAT6 will support a 250 MHz bandwidth for data. CAT7 standards have not been set yet. CAT7's proposed data rating is a 700 MHz bandwidth.

With ever-increasing speeds available for local data networks, fiber may eventually replace unshielded twisted pair altogether. See the section on local area network (LAN) wiring in Chapter 3 for more detail on unshielded twisted pair.

Coaxial Cable

Another type of transmission media is coaxial cable (coax). This cable can support much higher bandwidth than traditional twisted pair (TP). The coax inner conductor contains a solid copper wire, and the outer conductive wrapping serves as a shield. Any noise or interference from the surroundings is absorbed by the wrapping and sent to ground. Coax allows for the transmission of high frequencies but is less susceptible to noise than twisted pair (TP). On the negative side, coax is slightly more expensive than twisted pair, and it is often difficult to work with because of its size and rigidity. However, much higher data rates (in the order of several hundred Mbps) are achievable with this type of media (Figure 1.7).

Most cable television systems use coaxial cable to connect their customers (Table 1.3).

Table 1.3 Coaxial Cable Specifications

Type	Ohms	Loss at 100 MHz	Loss at 400 MHz	Center Conductor	Description (use)
RG-6	75 Ω	2.2 dB	4.4 dB	18 gauge	TV, cable TV, satellite, VCR
RG6QS	75 Ω	2.2 dB	4.4 dB	18 gauge	Same as RG-6 + increased shielding
RG-58	50 Ω	4.5 dB	10 dB	20 gauge	CB, ham radio, shortwave applications
RG-8	50 Ω	1.9 dB	4.1 dB	13 gauge	CB, ham radio, scanner

Fiber Optic Cable

Fiber optics is one of the major developments in telecommunications over the last several years. It provides an extensive bandwidth with virtually error-free telecommunications. As a result, the world has an ample supply of quality telecommunications circuits at a low cost.

Fiber optics' bandwidth is almost limitless. Billions of digital bits can be transmitted per second along an optical fiber link. This bandwidth contains enough capacity to handle thousands of telephone calls. Although fiber optic cable is still more expensive than standard copper wire, it is preferred for today's high-speed data communications. Another feature of fiber optic cable is the elimination of twisted pair cable problems, specifically, crosstalk and electromagnetic interference (EMI).

Lightwave communications has one major drawback. Its enormous bandwidth makes it vulnerable. A single cable break can disrupt telecommunications to a major area or an entire city. Therefore, it is desirable to have a backup path available. The backup fiber must travel a different path from the main fiber. Fiber diversity and automatic backup of fiber routes will prevent a service disaster when the main route is lost.

Fiber optic cable is composed of a transmitting laser and receiving laser (for short cables a less expensive LED can be used). The transmitting laser sends light signals through the fiber to the receiving laser. For example, when the light is on, a "1" may be received. Conversely, when the light is off, a "0" may be received (Figure 1.7). It is important to note that a fiber carries signals in one direction only. Therefore, fibers are used in pairs for bidirectional transmission.

The advantages of optical fiber are many:

- Good for high bandwidth applications
- Not subject to electrical interference
- Crosstalk between fibers nonexistent
- Lightweight construction
- Small physical size
- Minimal loss of signal strength
- Most secure medium—difficult to tap and does not radiate electrical signals

Two types of fiber are used. One type is called single-mode fiber, and the other is called multimode fiber. Single-mode fiber uses a glass core of 5 microns. This core size allows the light beam to travel with minimal reflection from the cladding. With this core size, a concentrated light source is used. Telephone and cable television networks install millions of strands of single-mode fiber every year.

Multimode fiber provides higher bandwidth over long distances. Light waves are broken up into numerous paths as they travel through the cable's core. Multimode fiber uses glass cores of 50, 62.5, and 100 micrometers. However, in longer cable runs, multiple paths of light can cause signal distortion. This distortion can result in unclear and incomplete data transmission.

Single-mode fiber costs more than multimode; however, single-mode fiber provides a higher transmission rate. Single-mode fiber also provides up to 50 times more distance than multimode fiber. Remember that when single-mode fiber is used, only one lightwave can be transmitted at a time. The small core and single lightwave transmission of single-mode fiber virtually eliminates any distortion, providing the highest transmission speed of any fiber cable.

It is always desirable to place a minimum number of repeaters along a network because of right-of-way and maintenance issues. Fiber has the least amount of attenuation of all guided transmission systems, making it the most desirable because of its minimal use of repeaters.

TIMETABLE OF HISTORIC FIBER OPTIC TRANSMISSION EVENTS

1880	Alexander Graham Bell invents the photophone: a device that modulates a light beam from the sun.
1960	Laser development begins.
1967	Pure glass fiber is developed.
1970	Donald Keck (Corning Corp.) shoots a beam of light through a modified glass strand and records information at the opposite end.
1977	First commercial fiber optics system operates at 45 Mbps.
1978	Corning Corp. opens the first fiber manufacturing facility.
1988	AT&T installs the first trans-Atlantic fiber (TAT-8), which can handle 40,000 simultaneous voice calls.
1997	Corning Corp. unloads its housewares glass division and produces fiber exclusively.

Today, fiber optic testers shine a light from one end of the cable. At the opposite end is a receiver calibrated to the strength of the light source. With this test, one can measure how much light intensity is getting to the other end of the cable.

A major controversy is whether fiber should be installed all the way to the desktop computer. Costs are dropping to the extent that fiber may soon replace all other guided transmission systems.

UNGUIDED TRANSMISSION MEDIA ....

Unguided media can be considered a transmission system through the air, water, or space. However, we will concentrate our studies on the air interface. In the United States, the Federal Communications Commission (FCC) divides the air frequency spectrum for our use. This government agency sets the standards for AM/FM radio, microwave radio, cellular telephone, etc. (Table 1.4).

Table 1.4 Unguided Frequency Spectrum

Name	Band Name	ITU Band	Frequency Range	Waveform Name
Very Low Frequency	VLF	4	3–30 kHz	Myriametric
Low Frequency	LF	5	30–300 kHz	Kilometric
Medium Frequency	MF	6	300–3000 kHz	Hectometric
High Frequency	HF	7	3–30 MHz	Decametric
Very High Frequency	VHF	8	30–300 MHz	Metric
Ultra High Frequency	UHF	9	300–3000 MHz	Decimetric
Super High Frequency	SHF	10	3–30 GHz	Centimetric
Extremely High Frequency	EHF (IR)	11 12	30–300 GHz 300–3000 GHz	Millimetric Decimillimetric

Amplitude-modulated radio (AM radio) and frequency-modulated radio (FM radio) were already discussed earlier in this chapter. Notice that FM radio has a much wider bandwidth than does AM radio (Figure 1.6). AM covers 550 kHz to 1600 kHz, or a bandwidth of 1.5 MHz (subtract the lowest from the highest). FM covers 90 MHz to 107 MHz, or a bandwidth of 17 MHz (subtract the lowest from the highest). That's over 10 times more bandwidth allocated for FM radio than for AM radio. Notice the difference in sound quality.

Let's now analyze microwave radio in more detail.

Traditional Microwave Services

Microwave technology is still an attractive method for obtaining wide bandwidth telecommunications in remote areas. Microwave radio was the primary transmission method of the 1950s, 60s, and 70s for telecommunications carriers.

Many of the microwave transmission facilities installed since this time are now idle. Optical fiber, because of its error-free superiority, replaced much of the microwave network in the 1980s and 90s.

Microwave technology is still an extremely attractive transmission method for today's bandwidth-hungry applications. Here are some examples:

- Temporarily restoring damaged wiring or circuits
- Crossing obstacles such as mountains, highways, or rivers
- Transmitting video from remote to studio
- Bypassing local telephone network
- Providing diverse backup wiring to protect against main circuit failure

Microwave frequencies have a frequency spectrum between 1 and 40 GHz. Although this is an extremely large bandwidth, microwave paths are very congested in metropolitan areas. It should be noted that frequencies higher than 30 MHz travel in a straight line. Transmissions at these higher frequencies are ideal for short-distance, line-of-sight telecommunications. Microwave signals are affected by the following:

- Rain attenuation: Occurs when raindrops scatter the signal
- Diffraction: Occurs as a result of the landscape
- Free space loss: Occurs as the signal travels through the atmosphere
- Reflections: Occurs as the signal takes multiple paths that arrive out of phase and can cause the signal to fade
- Interference: Occurs when other frequency transmissions interfere with the main signal

Microwave systems use amplitude- or frequency-modulated transmission signals, which are connected between two antennas in a line-of-sight configuration. At the high end of the microwave frequency spectrum, antennas may be facing each other within 1 mile. At the lower end of the microwave spectrum, antennas may be spaced 30 miles or more from each other (Figure 1.8).

Microwave frequencies act similarly to light waves. They can be focused by a parabolic dish or horn. The diameter of these dishes is from 2 feet for short hops to 100 feet for longer hops. At lower frequencies, antennas can be fed from coaxial cable. At higher frequencies, waveguides must be used. A waveguide is a rectangular device designed for the transmission frequency range.

Microwave radio can be considered as a viable solution when the following factors are considered:

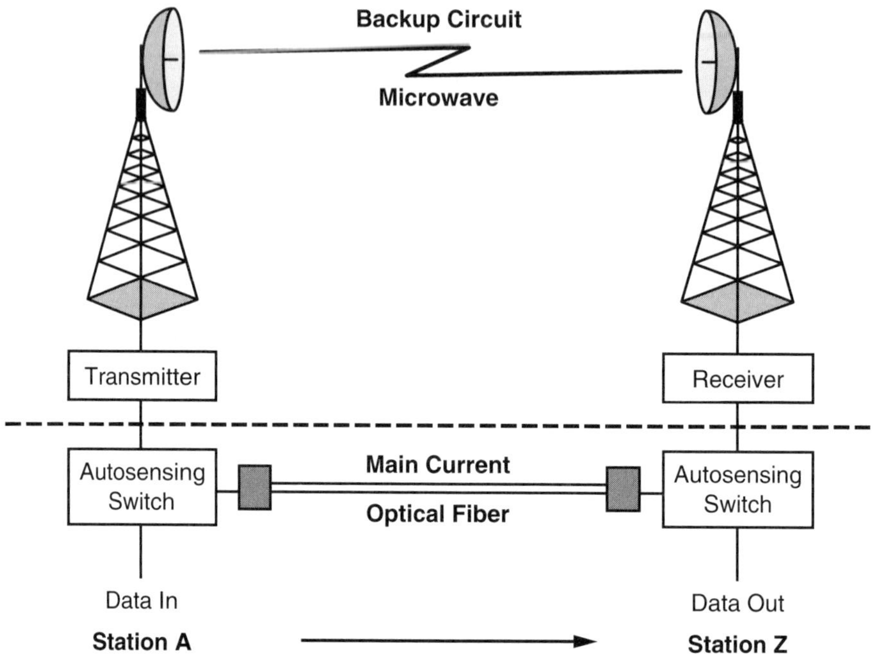

Figure 1.8
Microwave transmission used as a redundant backup.

- Cost effectiveness: Can microwave save money over local telephone solutions?
- Bandwidth limitations: Can microwave provide enough bandwidth for future needs?
- Line of sight: Is there a line-of-sight between microwave dish terminals? Microwave repeaters may be necessary if line-of-sight is not available.
- Temporary connection: Is a temporary connection needed for a remote sight? If this is true, then a microwave solution may provide quick and easy access.

In today's world, microwaves can be used with remote news-gathering trucks that point a microwave dish back toward a television studio. Remote stations can be quickly set up, configured, and placed online for live media coverage.

Let's take microwaves one step further and examine what happens when we place a microwave relay tower in space. This is an example of a telecommunications satellite system.

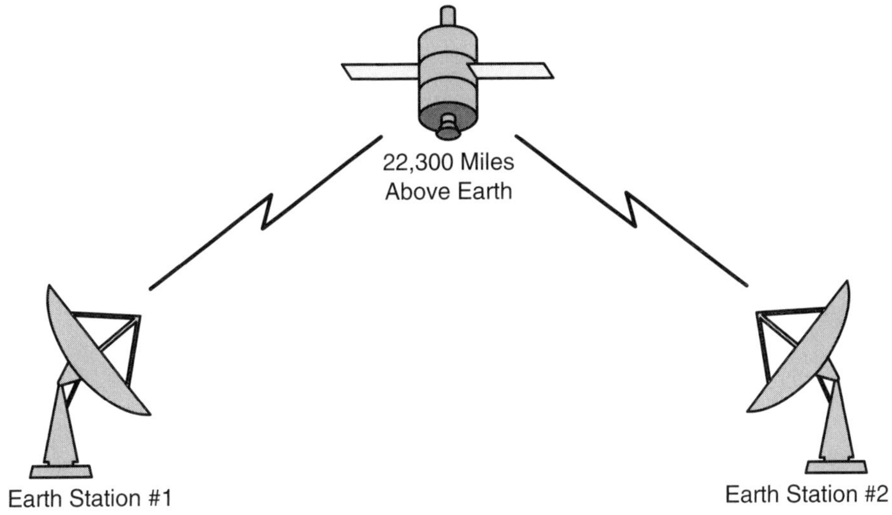

Figure 1.9
Traditional geosynchronous satellite service.

Traditional Satellite Services

Telecommunications technologies have evolved over the years from land-based copper wiring to satellite and microwave communications and then to optical fiber. Satellite has several advantages over fiber. One, fiber cannot economically reach remote locations. Two, it cannot broadcast a signal to all parts of a continent. Three, it cannot receive and transmit a signal when moving.

The first satellites carried several voice circuits in a low earth orbit. Earth stations (antennas on the ground) tracked the satellite as it moved and before it disappeared over the horizon until the next satellite appeared. This proved to be inefficient because a chain of satellites would be needed to provide constant communications.

It was soon realized that communications satellites would be required to orbit above the equator at a geosynchronous altitude of 22,300 miles above the earth. This orbit covers both the northern and southern hemispheres. The coverage area is sometimes referred to as the satellite footprint. At geosynchronous orbit, satellites travel at the same speed of the earth's rotation. Therefore, to us it appears as if a satellite is stationary or at a fixed position (in relation to the earth) (Figure 1.9).

Table 1.5 Satellite Band Designation

Satellite Band	Frequency
L-Band	1–2 GHz
S-Band	2–4 GHz
C-Band	4–8 GHz
X-Band	8–12 GHz
Ku-Band	12–18 GHz
Ka-Band	18–30 GHz

Frequency bands for satellite transmission have been divided into alpha characters (this was done for World War II top-secret information). In the early days of satellite telecommunications, the C-band was used. Ku and Ka bands became popular after the C-band ran out of space (Table 1.5).

Satellites have several drawbacks, which may include the following:

- Delay from earth station to satellite is ¼ second, creating an echo effect
- Frequencies within certain bands are crowded with a potential for interference
- Rain absorption affects signal loss, particularly at higher frequencies

Delay is a big problem with satellite voice transmission. The ¼-second delay between two earth stations can create a difficult voice conversation. In the 1960s and 70s, most international calls were handled by this method of transmission.

Transmission from an earth station to a satellite is called an uplink. Conversely, transmission from a satellite to an earth station is referred to as a downlink. Earth stations vary from a simple receive-only dish to an elaborate bidirectional dish more than 30 meters in diameter.

Satellites are powered by solar batteries. The power output and capacity obtained from these batteries can limit the life span of a satellite. A satellite also contains telemetry equipment to monitor its position. Telemetry also monitors signal strength and makes adjustments when necessary.

For a period of 10 minutes a day during the spring and fall equinoxes, the sun is positioned directly behind a satellite and directs a high level of radiation toward the earth station, making the satellite circuits unusable. This is referred to as a sun outage. Either a backup system or another satellite must be used during these periods of unusability.

Recently, traditional satellite services have proved to be very effective for one-way television broadcast transmission. An international satellite provides much of the overseas television programming we see on cable television.

V-SAT Services

Very small aperture terminal (V-SAT) is named for the size of the remote transmitting antennas. These antennas measure approximately six feet in diameter or less, which makes them easy to install. Generally, a V-SAT system contains a larger earth station hub with several remote stations or V-SAT terminals (Figure 1.10).

This service is designed as a private network used to bypass local telephone providers. The primary application for V-SAT is data transmission for remote locations. However, voice and video can be transmitted also. As the number of remote locations increases, the more cost effective the service becomes.

Timetable of Historic Satellite Transmission Events

1945–1957	United States (Werner von Braun) works on a satellite design.
1957	USSR shocks the world by orbiting Sputnik satellite.
1962	AT&T orbits the Telstar satellite, which becomes the first active telecommunication satellite in a low earth orbit (LEO).
1963	Hughes Corp. launches Syncom1 in a geosynchronous orbit. Syncom1 explodes five hours after launch. Syncom2 (launched later in the year) then becomes the first geosynchronous satellite. Syncom2 later provided extensive telecommunications during the war in Vietnam. Syncom3 (launched in 1964) provided a television broadcast of the Tokyo Olympics.
1963	United States forms COMSAT to represent the satellite interests of the country.
1964	INTELSAT forms as a not-for-profit cooperative of 140 member nations.
1965	INTELSAT1 is launched, providing 2400 voice circuits between New York and Europe.
1988	Hughes Corp. launches OPTUS-B, a large-capacity body-stabilized satellite.
1993	DBS2 is launched for direct broadcast television.
1995	DBS3 is launched for direct broadcast television.

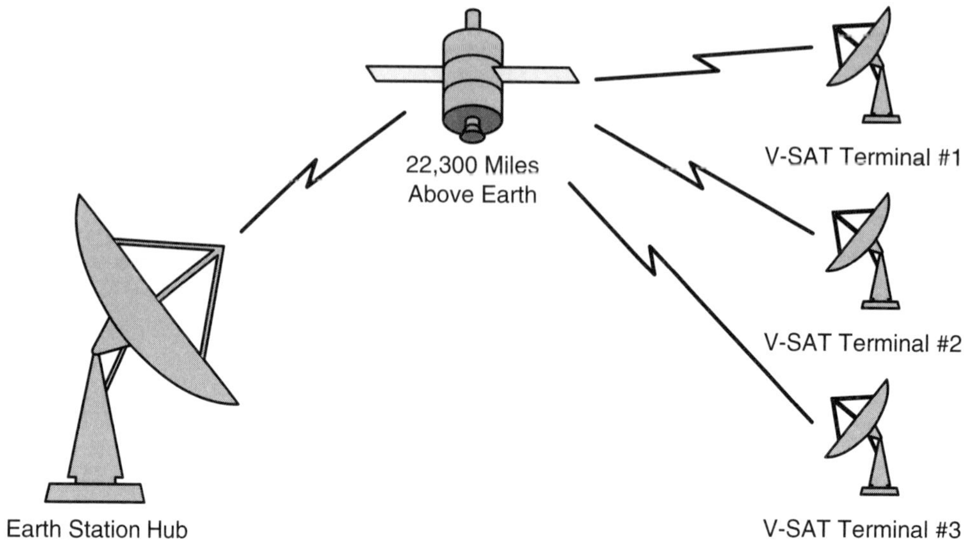

Figure 1.10
V-SAT service.

A good example of this type of network is a corporate office transmitting information to its regional offices. Typically, many gas stations have V-SAT antennas on their canopy rooftops.

Infrared

Infrared telecommunications services are gaining in popularity because they can handle high-speed transmissions without FCC licensing. Infrared frequencies are higher than visible light frequencies; therefore, they cannot be seen. This fiberless type of transmission can be referred to as unguided optical transmission.

There are two basic problems with infrared transmission. One, it cannot travel around objects such as walls. Two, it is subject to "rain fade." It is basically a line-of-sight transmission system.

A good example of a high-speed infrared application is two buildings within a mile of each other with a requirement to transmit and receive high-speed data. Each infrared device would be set on the roof of each building and aimed directly at the other. Each infrared transmission device contains a laser with the ability to exchange high-speed data rates (up to 155 Mbps digitally).

Note that the distance between transmitter and receiver should never exceed 5 miles and that the setup costs to create an infrared transmission system are relatively low.

CONCLUSIONS

This chapter serves as our basis for telecommunications transmission. We discovered that a circuit works by means of a conductor or wire, and we used Ohm's Law to check the voltage, current, and resistance. We learned that the decibel measures the ratio of power both into and out of a circuit. A simple telecommunications circuit was created in accordance with these concepts.

Unfortunately, many telecommunications textbooks do not include basic electronics concepts. Electronics fundamentals are the basis for all networks and circuits. It is an important area of study that is too often overlooked.

We discovered the differences between amplitude-modulated signals and frequency-modulated signals. The advantages and disadvantages of each were also discussed. The analog world containing frequency-modulated and amplitude-modulated systems is still with us. The best examples are AM and FM radio. These technologies show no signs of fading away soon.

Twisted pair copper, coaxial cable, and fiber optic comparisons were made. It was obvious that optical fiber was the winner, providing as it does a virtually error-free signal from end to end. As the cost of fiber continues to drop and splicing methods continue to improve, we can expect to see even more fiber in telecommunications networks, especially at the local end.

The last section of the chapter contained information on unguided systems, specifically microwave and satellite transmission. These methods of transmission were used extensively before optical fiber became the obvious choice. Microwave and satellite transmission systems still have much life left and will still play a major role in telecommunication networks of the future.

Chapter 2 closely analyzes the standard telephone network. Specifically, telephone switching and signaling capabilities are emphasized.

2 Traditional Telephone Services

Now that we understand the basics of transmission, let's take a closer look at one of the most universally accepted devices in the world: the telephone. The telephone began with Alexander Graham Bell's invention for the deaf and grew into a standard device that is essential for both businesses and consumers alike.

Traditional telephone services are also known as the public switched telephone network (PSTN) or plain old telephone service (POTS). Both PSTN and POTS are the acronyms for the standard telephone technology we all grew up with. There have been changes during the last 100 years, but relatively few compared with other technologies that have evolved during our lifetime.

Most of these recent changes are the direct result of digital transmission and high-capacity fiber.

This chapter is divided into seven sections.

In the first section, we review the invention of the telephone itself.

In the second section, we take a brief look at telephone history and examine why and how we got where we are today. Policy and regulation in the telecommunications industry at times have been very turbulent. We review some of the decisions leading up to the breakup of AT&T and the most recent Telecommunications Act of 1996.

The third section analyzes the telephone device itself. The microphone, speaker, switchhook, dialer, and ringer are all components of this simple but elegant device. Also, we will build a simple telephone circuit, based upon our knowledge of circuit theory from Chapter 1.

The fourth section covers telephone network access. Terminology such as local area transport access (LATA), LEC, CLEC, IXC, etc. is discussed. Circuits involving the central office (CO) switch, local loops, and demarcations are discussed.

In the fifth section, we cover telephone switching. In particular, we cover basic switching and circuit switched services in some detail. We also discuss blocked calls and the need for capacity planning.

The sixth section examines telephone signaling and the differences between in-band and out-of-band signaling. Signaling system 7 (SS7) basic architecture and features are examined also.

The seventh and final section of the chapter takes a closer look at data services using PSTN. The functions of PSTN dial-up modems and ISDN modems for digital services are analyzed. We take a closer look at modems and see why they are necessary in today's digital world.

Bringgggggg... Bringgggggg... Bringgggggg...

EARLY TELEPHONE COMMUNICATIONS................

Until recently, telephone history in the United States was that of one entity, the Bell Telephone System. The Bell Telephone System was named after its inventor, Alexander Graham Bell, who received the first patent for a telephone in 1876.

During the period of monopolistic regulation (from the 1913 Kingsbury Commitment to the 1984 Telecommunications Act), there was a need for universal service. The importance of a telephone network could be ascertained by the number of its customers. During this period of intense regulation, the Bell System (AT&T) worked with state utility commissions and the Federal Communications Commission (FCC) to provide this coverage.

The cost of basic service was kept to a minimum, with revenues made up from other services such as business services and long distance services. In the 1970s the Bell System claimed that it lost about $7.00 a month on its basic service rates. Extension telephones, colored telephones, touch-tone services (starting from the 1970s), and coiled handset cords all provided additional revenues on top of the basic fee.

The Bell System comprised 22 Bell Operating Companies (BOCs) formed to provide local services. Bell Laboratories was the engineering and design arm of the Bell System. AT&T provided long distance services, and Western Electric provided the equipment for the network. Consumers leased their telephone, which was owned by the Bell System. This lease was included in the cost of basic service.

Figure 2.1
Early telephones.

The Invention of the Telephone

Analog voice telecommunications have been in existence for over 120 years with the start of Alexander Graham Bell's telephone invention. The first phone conversation from A. G. Bell to Dr. Watson (his assistant) was "Mr. Watson, come here; I want you." This actually was the first point-to-point voice telecommunication (Chapter 3). There was no network, and the bandwidth was limited to microphone, speaker, and wire technology available at the time (Figure 2.1).

In a bitter legal patent infringement, Bell eventually won the case against Elisha Gray. Gray's own phone design had arrived at the patent office only three hours after Bell's. Gray, however, worked with Enos Barton and formed Gray and Barton, which later became the Western Electric manufacturing company. Western Electric became the sole provider of telephone equipment and devices for many years.

Bell's invention of the telephone grew out of his interest into ways to improve the telegraph. Patent number 174,465, issued on March 7, 1876, covered the method of transmitting sounds telegraphically through an electrical current in a wire.

Bell first demonstrated his most famous invention in 1876 at the Centennial Exhibition in Philadelphia. He proved that the sound of the human voice could be reproduced electrically. This demonstration confirmed that speech patterns could be made to change an electrical current.

The world's first phone call crossed approximately 8 miles between two towns in Ontario, Canada. Soon thereafter, telephone technology was expanded to over 100 miles. The Bell Telephone Company was established in 1877 to bring telephone service to everyone. The company provided the foundation for today's telecommunications industry.

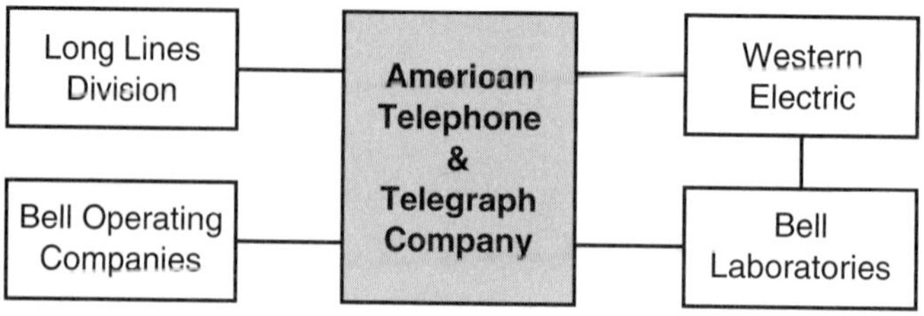

Figure 2.2
Bell System (AT&T).

After the invention of the telephone, Bell continued his experiments in telecommunications. One of his patents, issued in 1880, actually transmitted sound via a beam of light. This experiment, called the photophone, could be considered a precursor of today's optical fiber systems and the first wireless transmission of speech.

Telephone Policy and Regulation

In the early days, the analog telephone network was composed of small telephone companies that operated independently of each other. In 1893, over 5,000 of these independent companies were providing telephone service and selling telephone equipment. This led to competition among the independents, which in turn led to confusing allegiances of customers who could not be connected to clients of a competing phone company. Concerns over this lack of standardization and interconnection prompted government oversight of telephony in 1910 (Figure 2.2).

Kingsbury Commitment

During the early 1900s the Bell telephone network began acquiring independent telephone companies. Fearing a monopoly, the Justice Department filed its first antitrust suit against the Bell System (AT&T) in 1913. Bell System (AT&T) vice president Nathan Kingsbury then sent a letter to the Attorney General. This letter outlined Bell's promise to dispose of Western Union stock, cease acquisition of independents, and provide interconnection for all independent companies to the Bell System network.

The Justice Department ended its antitrust action on the basis of this letter, which is now known as the "Kingsbury Commitment." This case marks the beginning of the first nationwide telephone network in the United States.

Bell Telephone—A "Natural" Monopoly

The FCC endorsed the industry's structure, characterizing it as a "natural monopoly" in 1939. By the late 1940s, however, the Justice Department began to feel uneasy about the sheer magnitude of AT&T's empire.

Antitrust proceedings were initiated against AT&T in 1948. A 1956 consent decree was formed stating that the government would drop its lawsuit if AT&T would stay out of the computer industry.

Hush-a-Phone Decision

Under the guise of protecting its network from problems of unreliability and incompatibility, AT&T disallowed companies other than Western Electric from manufacturing equipment for the network. The telephone was therefore leased from AT&T. Consumers could not own any equipment connected to the telephone network.

The Hush-a-Phone device was created as a simple product that slipped over a telephone handset and gave the user some privacy. There were no electrical connections; it just cupped over the telephone's microphone.

This device had been advertised and sold in trade magazines for many years. AT&T had previously won several court cases against companies that sold paper advertising that could be attached to the phone. It is hard to believe that at one time you couldn't attach even a phone list to your telephone since the telephone was leased, not owned, by the consumer.

On appeal, the court held that the AT&T ban of the Hush-a-Phone attachment device "was neither justifiable nor reasonable" on the basis that the device emulated the natural cupping motion of one's hand.

Carterphone Decision

In 1968, the FCC ruled against AT&T's ban of an acoustic coupler connecting radiotelephones. Acoustic couplers were the predecessors of today's modems. A telephone handset was placed into two rubber cups in the acoustic coupler. The acoustic coupler modulated and demodulated message data, usually at a very low speed of about 300 baud.

A radio dispatch company known as Carter Electronics developed a simple device, which later became known as the Carterphone, that could pass messages directly through the PSTN. The FCC allowed this interconnection with the acoustic coupler because there was no electrical connection between the PSTN and the messaging system.

By allowing the connection of the Carterphone device, the government signaled that equipment not made by Western Electric, which was AT&T's manufacturing subsidiary, could be used in its network.

This was a milestone achievement. Up until 1968, AT&T's lawyers and engineers argued that any foreign devices (other than Western Electric) attached to the network could have severe implications. The Carterphone decision proved that these implications were not true.

MCI Decision

The AT&T long distance monopoly was challenged by Microwave Communications, Inc. (MCI) in the late 1960s. Customers looking to extend their networks could use MCI's microwave link between St. Louis and Chicago. This would require a connection into the PSTN, tying the customer's telephone systems together. The customer in either city could dial a local number and access another user in the other city.

AT&T filed a case in court that accused MCI of unlawfully infringing on its monopoly. The argument once again did not hold up in court. Both the courts and the FCC declared that it was in the public interest for MCI to provide this service.

Telecommunications Act of 1984

In 1975, the government filed another antitrust suit against AT&T, alleging unlawful monopoly. This case was litigated until Judge Greene issued the Modified Final Judgment in 1982, which later became known as the Telecommunications Act of 1984. This act called for the breakup of AT&T and allowed the new AT&T to compete in long distance services (Figure 2.3).

The Telecommunications Act of 1984 can be summarized as follows:

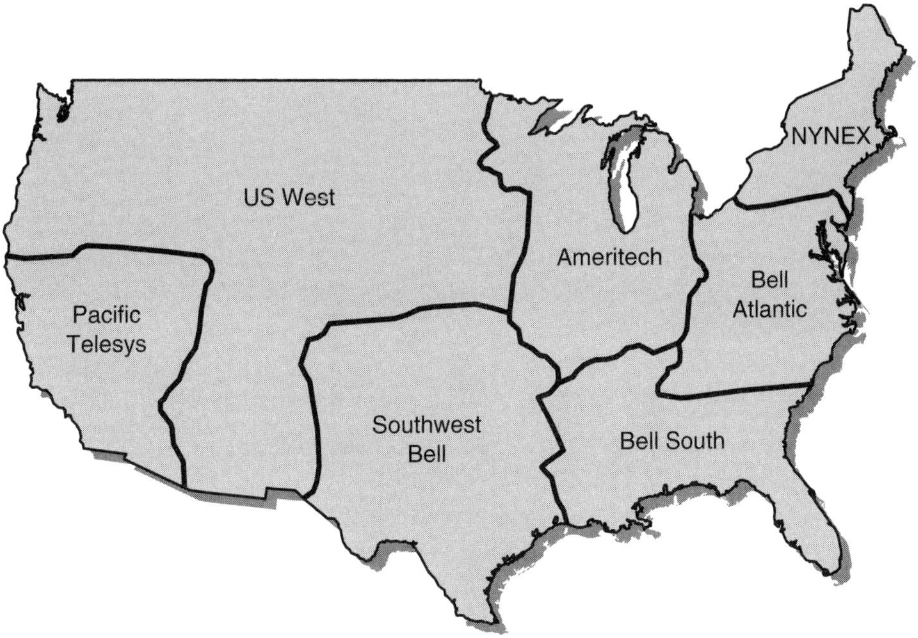

Figure 2.3
RBOCs in the United States (1984).

- Twenty-two Bell Operating Companies (BOCs) were spun off into seven Regional Bell Operating Companies (RBOCs) (also known as Local Exchange Carriers (LECs)).
- RBOCs would provide equal access to all long distance carriers or interexchange carriers (IXCs): AT&T, MCI, and Sprint.
- Bell Laboratories would be split in two, with one half going to the RBOCs (called Bellcore Inc.) and the other half to remain with AT&T.
- Western Electric would remain with AT&T but would no longer be the sole provider of telecommunications equipment for the PSTN.

The terms of this agreement prohibit the RBOCs from transporting long distance data outside of geographical boundaries called local access transport areas (LATAs). Similarly, IXCs would be prohibited from transporting local data from within LATA boundaries. The role of the Telecommunications Act of 1984 was to keep the RBOCs as local providers and the IXCs as long distance providers (Figure 2.4).

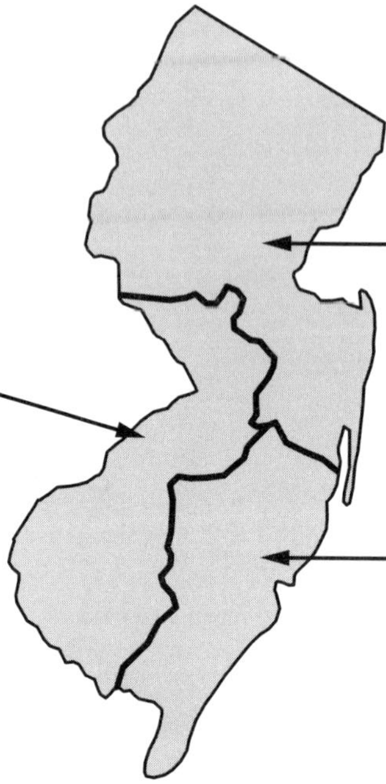

Figure 2.4
Local access transport area in New Jersey.

AT&T agreed with the Department of Justice to provide equal access to all IXCs. Equal access requires switching intelligence to designate a default long distance provider (that is, AT&T, MCI, Sprint).

In defense of AT&T, we can say that they achieved the goal of providing universal service, with the highest reliability available at the time. The company connected over one billion calls a day in 1984, at the time of its breakup. Yet, AT&T recognized that it was in its own interest to discontinue the monopoly in which it had become encased. Time would prove that this divestiture settlement was necessary in a growing telecommunications marketplace.

Telecommunications Act of 1996

The Telecommunications Act of 1996 was an attempt by the government to increase competition through deregulation. The deregulation has been gradual since all interested carriers must meet certain criteria. Under this act, local services could be offered by a variety of companies including IXCs (AT&T, MCI, Sprint), cable television companies, and even power utility companies. Similarly, the LECs (Verizon, Bell South, etc.) were allowed access to the long distance market, provided they could prove that there was enough competition in the local telecommunications market. For the first time since the Telecommunications Act of 1984, the LECs could potentially provide inter-LATA calls and the IXCs could provide intra-LATA calls.

TIMETABLE OF HISTORIC TELEPHONE NETWORK EVENTS

1877	First telephone line between Somerville, Massachusetts, and Boston.
1878	First telephone directory is issued.
1892	First automated mechanical switch (Strowger switch) is used.
1909	The Bell System (AT&T) and Western Union are intertwined.
1913	The Kingsbury agreement is created between AT&T and the U.S. government.
1925	Bell telephone laboratories are founded.
1935	First telephone call around the world is made.
1949	AT&T and Western Union introduce the famous model 500 black rotary telephone.
1950	Approximately 75% of all telephones are party lines.
1956	Hush-a-Phone court decision is issued.
1959	AT&T introduces microwave transmission.
1962	Comsat satellite service is formed.
1968	Carterphone decision allows interconnection of non-Western Union devices.
1969	MCI begins operations through the PSTN in the midwest.
1975	Fiber optic cable tests begin in Europe.
1984	AT&T is broken up.
1990	Coin phones now provide equal access.
1991	Caller-ID standards are in effect.
1996	ISDN standards are in effect.
1996	Telecommunications Act of 1996 is passed.

The FCC established guidelines, called the universal service fee (USF), for universal service. All telecommunications service providers are required to pay a small portion of their revenues to the USF to keep standard telephone service rates low.

Since the act became effective in 1998, there has been a series of telecommunications mergers and acquisitions. As of today, there are only four LECs: Verizon Corp. in the northeast, Bell South in the southeast, Qwest in the west, and SBC in the midwest. AT&T Corp. has purchased cable companies with the idea of establishing telephone and other services by cable. Worldcom Inc. purchased MCI Corp. and became MCI Worldcom, Inc. Sprint has also considered other purchases and mergers.

To consumer groups, the act has largely been a failure since competition, especially in the local market, has been slow and confusing. The long distance market now faces increased competition with declining revenues. The act has expected built-in points of failure, and most consumers have had few adverse effects.

International Telecommunications Union

The International Telecommunications Union (ITU), a United Nations agency, publishes worldwide standards for the telecommunications industry. The ITU was created in 1993 to replace the Consultative Committee on International Telephone and Telegraph (CCITT), which had origins over 100 years old.

ITU standards are nonbinding. They are based upon recommendations from technical committees. These committees are formed from telecommunications representatives throughout the globe. AT&T frequently bypassed the ITU/CCITT. For example, we will see in Chapter 3 how both European and North American digital carrier formats established in the 1960s are incompatible.

Standards and agreements are absolutely necessary in a telecommunications environment. This is the reason why the ITU is so important. However, it should be noted that standards may take several years to produce and many manufacturers want to be the first to market a new technology. A good example of this dilemma is the V.90 ITU standard for the 56-kbps modem. Two competing manufacturers developed products in advance of the V.90 standard to support 56-kbps transmission. Some of these modems were upgradable to the V.90 standard, but others were not. This caused incompatible products and lots of confusion (we discuss modem technology later in this chapter in "Modems").

The next time you call a long distance or international number, think about how much equipment and wires are needed to terminate the call. All of this is possible because of standards and cooperation between telecommunications carriers and countries.

At present, more than 2,500 recommendations are under evaluation by the ITU. The ITU is based in Geneva, Switzerland.

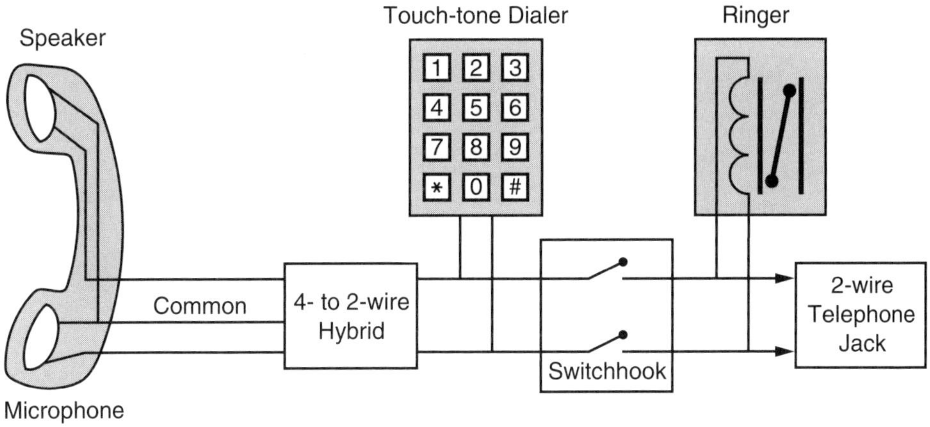

Figure 2.5
Basic telephone device.

The Telephone Terminal

Analog telephone end-terminal devices have evolved very little over the last 100 years. The device usually consists of a microphone, speaker, dialer, and on-hook/off-hook switch. It is one of the simplest electronic devices installed in the home. The signal from the telephone company hasn't changed in over 100 years. Therefore, the equipment hasn't changed much either. Let's take a closer look at this equipment (Figure 2.5).

The Telephone Handset

The telephone handset contains a microphone and speaker used to transmit and receive analog voice signals (Figure 2.6).

The function of a microphone is to convert sound waves into electrical energy. Traditionally, a carbon microphone was used. This type of microphone contained carbon granules sandwiched between two thin metal plates. The sound waves from a voice compress and decompress the granules, changing the resistance. We know from Ohm's Law that a change in resistance will affect the current flowing through the microphone.

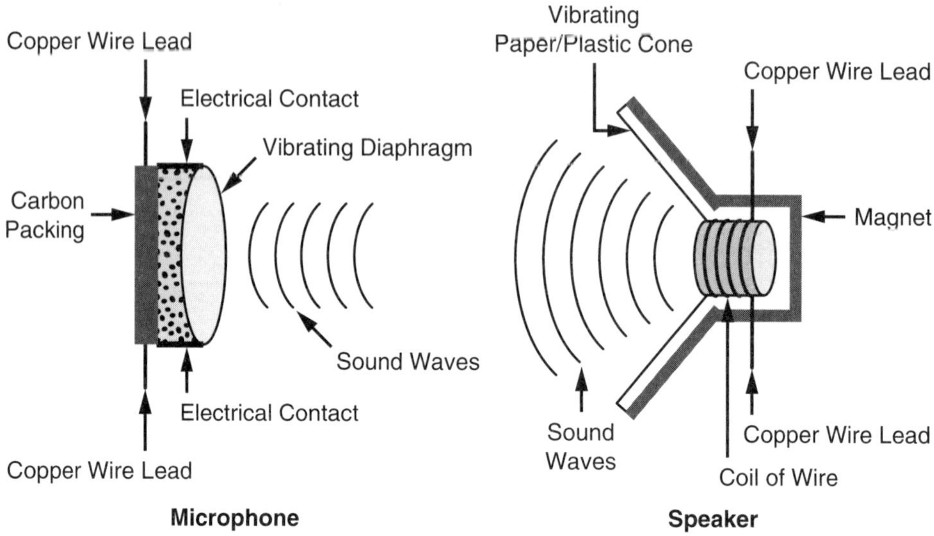

Figure 2.6
Microphone and speaker.

The varying current is an electrical representation of the analog voice. Therefore, we call this an analog signal. An analog signal continuously varies just like sound waves do.

Today's telephones use condenser microphones that are much smaller. These microphones use the principles of capacitance to regulate the current in accordance with the sound waves.

The function of a speaker is to convert electrical energy into sound waves (the opposite of a microphone). The speaker contains a magnet with a coil connected to a paper or polypropylene cone. The cone vibrates according to the electrical signal received in the wire. The cone's vibration changes the air pressure, which in turn produces audible sound waves.

Because of the 3 kHz narrow bandwidth used in telephone systems (see Chapter 1), the speaker is generally very small and is designed to reproduce these narrow bandwidths. Audio high-fidelity systems use much larger speakers capable of reproducing much larger bandwidths.

The telephone handset holds both the microphone and speaker in a housing that will keep the microphone close to the mouth and the speaker next to the ear.

Telephone Hybrid

The telephone hybrid is a device that converts four wires (two wires for the microphone and two wires for the speaker) into a two-wire system ready for telephone transmission. The hybrid device is also known as a duplex coil.

This device feeds back the sound of a voice from the microphone into the speaker. This feedback, called sidetone, allows the user to hear both the transmitted and received information through the speaker in the handset.

Telephone Dialers

Back in the early days of telephone transmission, the telephone did not have a dialer. Instead, an operator would speak with the call originator and decide how to route or "patch" the call. Of course, today we have sophisticated electronics and circuit-switching techniques to route calls to their destination within seconds. Much of the routing and switching available today starts with the telephone dialer (Figure 2.7).

The rotary dial generates pulses on the telephone line by opening and closing the circuit when the dial is turned clockwise and released. The dialer acts like a switch. This switch can pulse the circuit up to 10 times for the number 0. For example,

0 = ON-OFF-ON-OFF-ON-OFF-ON-OFF-ON-OFF-ON-OFF-ON-OFF-ON-OFF-ON-OFF-ON-OFF-ON-OFF-ON

The number of pulses (or circuit opening and closings) is determined by how far the dial is turned (1 is the shortest and 0 is the longest). When the dial is released, a spring forces the dialer back to its original position and the circuit pulses. Pulse dialing worked well with electromechanical switches, which were in use by AT&T prior to the 1970s.

The rotary dial generates pulses at a rate of 10 pulses a second. Each pulse lasts 1/20 of a second with 1/20 of a second between pulses. A phone with a keypad that has a dial setting generates an electronic pulse to simulate the old mechanical rotary dialer.

Dual-tone multifrequency (DTMF), or touch-tone signaling, works by sending tones via a keypad on the telephone line. The keypad is a tone generator with a 3 x 4 matrix. When a button on the keypad is pressed, two tones are generated with a dual-frequency combination. This is similar to the sound of two keys on a piano. For example, if you press the number "2," the frequency combination of row 1 and column 2 will result.

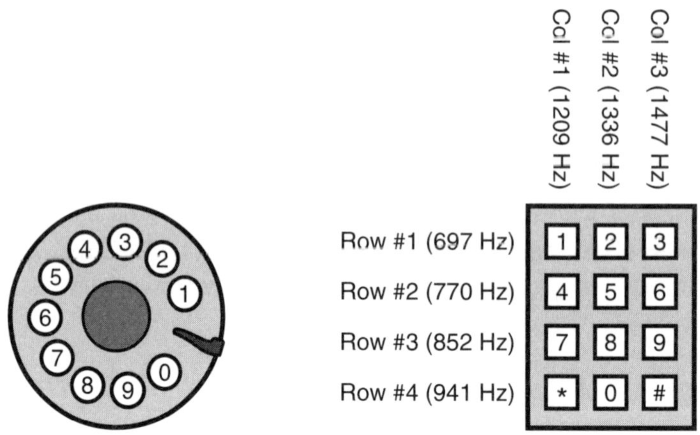

Rotary-pulse Dialing

Touch-tone Dialing
Dual-tone Multifrequency

Figure 2.7
Comparisons of pulse and touch-tone dialing.

Touch-tone dialing is considerably faster than pulse dialing. The average telephone number can be dialed 10 to 15 times faster with this dialer. A touch-tone dialer, or DTMF keypad, has replaced the pulse dial, or electromechanical switch dialer, in the United States.

Telephone Switchhook

The telephone switchhook is a basic switch that signals the telephone network that the telephone is on-hook (telephone handset is resting in its cradle and the telephone is not in use) or off-hook (telephone handset is not in its cradle and the telephone is in use).

The switch can open the circuit (current will not flow) or close the circuit (current will flow) and the phone is in use. You can also use the switchhook to simulate pulse dialing by opening and closing the circuit several times just like the old rotary phones discussed above.

Telephone Ringing

The telephone ringing device notifies the user that a call is incoming and someone is trying to reach the user. When the local telephone company sends an alternating ring signal, it is intermittent and is not sent constantly. Generally, ring timing is .2 seconds off with .4 seconds on, and so on. Ring timing varies from country to country.

The ringer equivalent number (REN) is a measurement of how much power a telephone can handle. RENs are used to determine how many telephones can be connected on a line before the telephones will not ring properly.

Audible ringing tone is the ringing that can be heard while the telephone is in use and can be heard when call-waiting features are used.

Another type of ringing is the distinctive ringing feature whereby different ringing tone patterns distinguish different numbers. For example, ring patterns for the main number may be .2 seconds off with .4 seconds on. A second number (using the same line) will have a ring pattern of .2 seconds off with .3 seconds on and .1 second off and .3 seconds on.

Ringers vary in size, shape, and sound. The old electromechanical "bell" ring will work on the same telephone circuit as the new electronic ringers that play a synthesized song.

Building a Telephone Circuit

The telephone is a simple electronic device that has been ingrained in American culture. It is so simple; you can create an intercom by using two or more telephones. A 12-volt battery is used to simulate power from the phone company. Standard 24-gauge telephone wire (or larger) can be used to connect the telephones (Figure 2.8).

The standard telephone connection is two wires for each telephone. Connect the wires from one phone to another, creating a circuit. The only device to add is the battery, which should be placed in the circuit loop.

Now, pick up the handset from each telephone and go off-hook. This action allows the electrons to flow within the circuit. You should now be able to hear your transmitted voice and the received voice through the handset speaker without any connection to the telephone company!

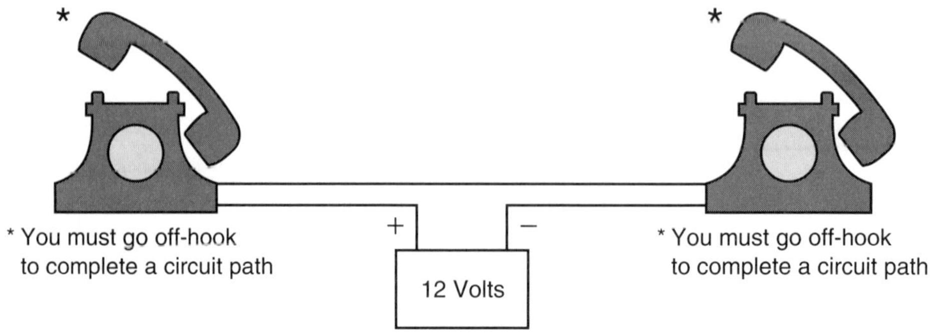

Figure 2.8
Building a telephone circuit.

Notice that the DTMF dialer will produce tones. Without switching equipment provided by the local telephone company, these tones are useless. Also, the ringing device within the telephone cannot work without a high voltage battery (usually about 90 volts). Of course, the local telephone company provides the ringer voltage when someone is trying to speak with you through a telephone connection.

This small experiment proves how simple the telephone device really is.

TELEPHONE NETWORK ACCESS............................

Each telephone must be connected to the network to provide voice telecommunications. This connection between the telephone company and the telephone is called the 2-wire local loop. Each local loop is extended to a telephone switching office called a central office (CO). The CO provides network access as well as a telephone battery. Tandem switching offices provide access to remote switching offices and to other carriers.

Three types of telephone calls exist:

- Local Calls—Calls using the same CO for both the call originator and the receiving end are called local calls. These calls are free. The cost of providing this service is included within the standard monthly rate (approximately $20) for telephone service. These are intra-LATA calls within a specific geographic boundary.

- Local Toll Calls—Calls within the same state and also within the same LATA (intra-LATA) using more than one CO are called local toll calls. One or more tandem offices may be involved to provide this service. These calls are billed as a per-minute charge. After the Telecommunications Act of 1996, IXC carriers (carriers other than an LEC) could handle these calls in some states (as determined by the local utility commission). The telephone company also uses trunks or multiplexed high-speed lines between switching centers. For example, after 1998, AT&T could handle local calls in New Jersey.
- Long Distance Calls—Calls between states. Interstate calls are also inter-LATA calls. In most cases, tandem offices are involved in the handoff to the IXC. High-speed trunks are also used between switching centers. After the Telecommunications Act of 1996, LEC carriers (carriers other than an IXC) could handle these calls in some states (as determined by the local utility commission). For example, after 1998, Verizon could handle long distance calls in New York.

The important thing to realize is that an IXC handling a long distance call must also pay access charges to each LEC (originating and receiving LEC). These charges eat into the IXC's profits. This is one of the reasons why the Telecommunications Act of 1996 was so important to the IXCs. Their profitability was significantly reduced by access charges (Figure 2.9).

Central Office Switching

Each geographical area contains a central office that provides network access for telephones in a specific region. The size of the region depends upon the population of the area. In the cities, the central office may serve a smaller geographic region than in a rural area.

The central office acts as a hub for all of the wires and cables connected to telephone poles, which ultimately connect to the individual telephone devices themselves.

The central office provides a telephone battery, which ranges anywhere from 8 to 12 volts (depending upon the distance of the line). Physical batteries or wet cells are used in case of power failure. If the electric utility can no longer provide service, telephones can still operate. The telephone batteries will operate for a limited time, generally a few hours. Generators provide electricity for lines and equipment before the batteries die out.

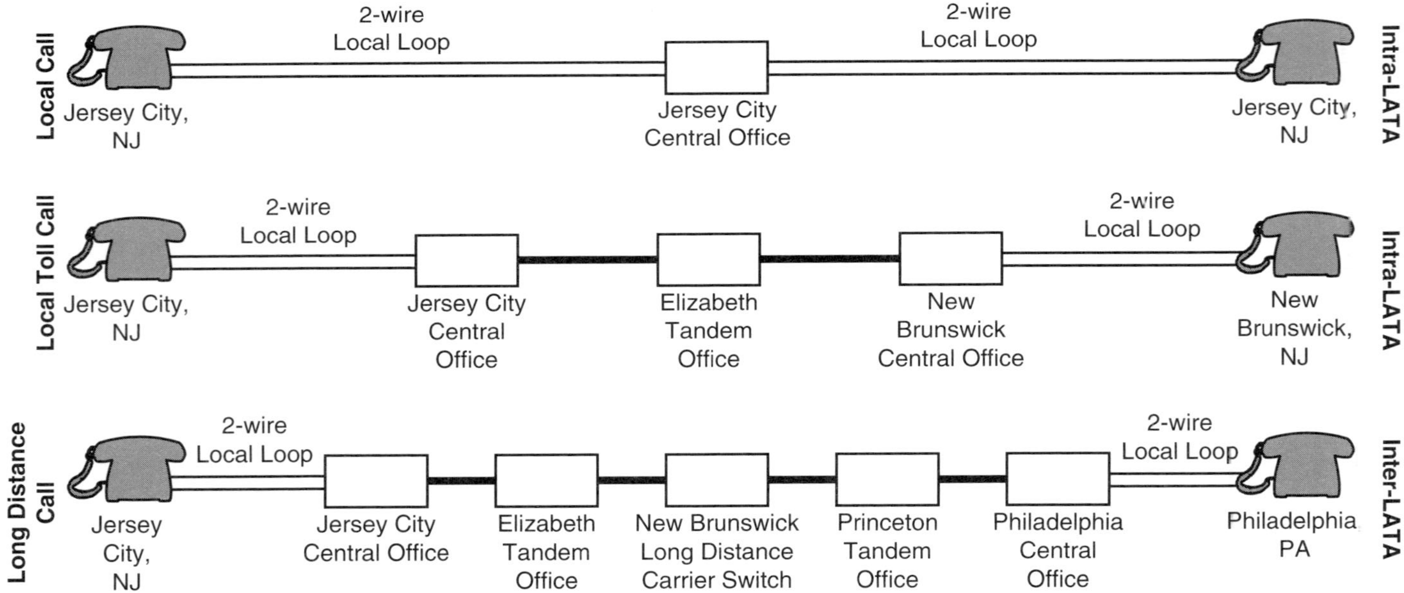

Figure 2.9
Telephone network access.

Wiring from telephone poles and underground cable are terminated in a rack called a main distribution frame (MDF) in the central office. From the frame, a fuse or loading coil protects the equipment from a short-circuit condition. Short circuits can be caused by excessive water or bare copper wires touching each other.

The switching equipment in each central office will attempt to route the call to its final destination. This equipment also provides signaling for the network, as discussed later in this chapter.

Tandem Office Switching

Other telephone switching offices in the network can be labeled toll offices or tandem offices. These switching centers are designed to handle long distance voice calls or local toll calls through high-capacity trunks. Tandem switches are similar to central office switches except that a tandem switch has few, if any, local loop terminations.

The tandem switch interfaces with other tandem switches for local toll or long distance telephone access. Tandem switches have some intelligence. They should be able to sense congestion and overload conditions and thereby reroute calls.

An effective tandem switch should provide the tools necessary to control and monitor the network. The switch needs to provide information about trunk status and performance. We will discuss more about trunk performance and call blocking later in this chapter.

Telephone Local Loops

The 2-wire local loop comprises cables, telephone poles, and conduit that connect the central office switch to the telephone end-terminal device. The local loop is the weakest link of the network. That is not to say that the local loop does not perform well. It just happens to be the portion of the network with the least available bandwidth. The local loop is sometimes referred to as "the last mile."

The local loop generally consists of twisted pair copper wire enclosed in bundles with other wires. In the cities, much of this wire is underground. In the country or suburbs, most of the local loop wire bundles are on telephone poles.

Because of the enormous cost regarding access from the central office to an end-terminal device, the local loop hasn't changed much. Local municipalities require right of way and permits to dig up streets, and such access can be expensive. For example, if LECs were to replace all of the copper local loops with fiber, the work would take over 10 years at a cost the LECs could not afford. The problem with replacement is simple economics.

The local loop is certainly able to support the 3 kHz bandwidth required for telephone service. However, in this bandwidth-hungry information age, 3 kHz is very limited. DSL attempts to better utilize the local loop bandwidth, which can sometimes approach 1 MHz or more. We discuss more about DSL in Chapter 5.

Remember, the central office provides a battery from 8 to 12 volts. Longer local loops will have a lower voltage due to resistance. However, the telephone end-terminal device will work well at all of these voltages. The ringing voltage is 90 volts (with 60 interruptions per minute) supplied by the central office. The local loop will support this ringing voltage.

Telephone Trunks

Trunk or backbone circuits are the high-capacity multiplexed circuits needed to provide network access. We've already discussed how a tandem switch can be used to interface with these trunks, leaving the central office to provide the interface for the end terminal.

In the telephone network, trunk circuits are almost always fiber because of their high capacity. If a major outage occurs in one or several trunks, the telephone supplier uses switching techniques to avoid trunks that are out of service.

Redundancy

Redundancy is simply a backup circuit for a main line or trunk. High-capacity fiber lines almost always require a backup circuit. Ideally, redundant lines should follow a separate physical transmission path. For example, let's say that a main fiber line runs from 14th Street and 7th Avenue to 42nd Street and 7th Avenue in New York City. It is desirable to find another redundant fiber path. If a spare fiber can be found along 8th Avenue and still terminate at both 14th Street and at 42nd Street, great!

When I think of redundancy I always think of a fiber that was cut on a Route 21 overpass in Newark during the mid-1990s. The entire Newark airport was shut down for several hours because of this. There were no redundant fiber lines available then. I can virtually assure you that Newark airport redundant lines exist today!

Demarcation

The responsibility of the local loop ends at the demarcation point. This demarcation point is usually the telephone box outside a residence but not within it. Inside wiring and the telephone device itself are the owner's responsibility.

In other words, if there is a problem with the telephone device or wiring, the local telephone company will charge for the replacement or repair because the problem arises in the owner-maintained lines and equipment. If the problem occurs outside the home, the local telephone company must perform the replacement or repair. Outside the home would be considered the local loop.

The local telephone company may agree to maintain the wiring in a home for a fee generally less than $2.00 per month. The local telephone company may also elect to provide a leasing arrangement for the end-terminal telephone device.

A demarcation point is also used for businesses. For small businesses, the rules are the same as for residences. For larger businesses, the telephone company's demarcation point frequently extends into an equipment room or floor of the building.

Demarcation points exist not only in the local environment but also everywhere in the network. For example, an LEC would hand off the telephone voice transmission to an IXC for long distance access. A demarcation point can exist on a switching office wiring frame, One side of the wiring frame would be for the LEC, and the other side of the wiring frame would be for the IXC.

TELEPHONE SWITCHING

Where would we be without today's fast switching techniques? Let's review and analyze the types of switching methods from the early days of the PSTN to today's switches that use SS7 services.

Basic Switching

Switching is an essential part of any telecommunications network. The most famous switched network in the world is the PSTN. Without automated switching in the PSTN, we would rely on manual switchboard operators to do the job. It is estimated that half the U. S. adult population would need to be a telephone switchboard operator if today's automated switching devices were not available.

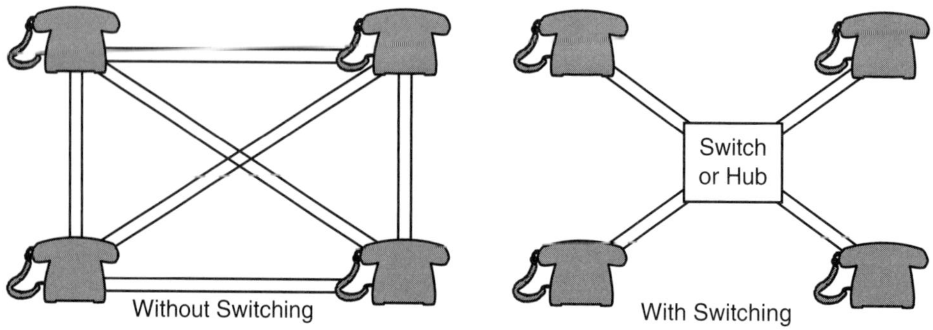

Figure 2.10
Basic switching.

Manual switching was fine when labor was inexpensive and telephone service was limited to a few wealthy individuals and businesses. Switchboard operators would monitor lamps to signal whether the receiver was on- or off-hook. A patch cord (cable with two wire connectors on each end) would be used to make the physical connection. Manual switching soon outgrew its usefulness as the telephone system grew rapidly.

To understand basic switching, let's say we have four phones that need to be interconnected with each other. We can install a pair of wires between each originating phone and each terminating phone. Connecting a pair of wires to the originating side and also connecting the same set of wires on the terminating side will establish a telephone call. If we need more than four phones, many, many more wires will be needed (Figure 2.10).

In the telephone network a local switch behaves as a hub (star configuration). Fewer wires are needed; therefore, the network is easier to troubleshoot. Also, when one telephone station is lost because of an open condition or short circuit, all of the other telephones connected to the switch are operational. We discuss more about the concept of a star or hub configuration in Chapter 3.

Numbering Plan

The 10-digit numbering plan, which consists of a 3-digit area code, a 3-digit exchange (or central office number), and a 4-digit number, was modified in 1995 when the area code structure of the number changed. The second area-code digit could now be a

TIMETABLE OF HISTORIC SWITCHING EVENTS

1877	First telephone line between Somerville, Massachusetts, and Boston uses manual switching (operators).
1891	Almon Strowger, a Kansas City undertaker, suspects that manual switchboard operators are routing calls intended for his business to the competition. He designs an electromechanical switch called the Strowger switch.
1938	AT&T introduces mechanical cross-bar central office switching.
1960	AT&T introduces the first electronic central office switch.
1965	AT&T introduces stored program electronic switching.
1970	AT&T introduces its ESS#2 electronic switch.

number from 0 to 9. Before this change, a second area-code digit could only be a 0 or a 1. This change was necessary because of the proliferation of cell phones, faxes, pagers, etc. As old area codes were filled, the telephone area could be split by use of both the new and old area codes.

The 800 numbers are used for reverse charging services. These services have separate pricing plans and can include intrastate 800, interstate 800, or both. All 800 numbers are stored in a nationwide database. When you dial an 800 number, the LEC determines which IXC handles the call for that number and attempts to terminate the call by using that IXC.

The 800 numbering was also modified in 1995, when 888 numbers were made available. The supply of 888 numbers lasted only 4 years. The code 877 was assigned in 1998 to supplement both the 800 and 888 numbering plans.

International dialing allows each country the freedom to provide its own numbering structure. However, the prefix of 011 must be used to tell the switch that the originating user intends to dial an international call. A country code follows the 011 code. Each country has its own code. The actual number (up to 11 digits) follows the country code. In 1997 these standards, adopted by the ITU, were expanded to accommodate 15 digits.

The 900 services are for businesses that want to play a prerecorded announcement or advertisement. Public opinion services can use 900 services also. Normally, the caller pays for the call. However, the business or sponsor may elect to pay for the call. Because of the content and high costs associated with 900 services, the user may elect to have the central office block these calls. The LEC can provide this blocked service free of charge. We discuss more about blocked calls later in this chapter, in "Blocked Calls."

Circuit-Switched Services

Circuit-switched services can also be called dial-up services. Our Plain Old Telephone Service (POTS) is the best example of a switched service. A temporary connection is created once call setup is complete. Since the circuit is created at the time of call setup, the circuit configuration is based upon network demand at the time.

It is important to note that switched services are not limited to the telephone network. Switched video services such as video-conferencing centers are an example. High-speed data switched services are also available.

Central offices handle thousands of telephone conversations daily. With the advent of electronic switching in the 1970s, the central office switch is noise free and more reliable.

What about the service itself? Well, the basics of switched services start with the touch tone dialer or dial pad. Let's say you want to dial a long distance call from New York City to your business associate in Los Angeles. Different scenarios depend on the condition of the network at the time of call setup (Figure 2.11).

The first switched-circuit routing scenario shows the central office (CO) and the IXC switch for both the originating city and the terminating city. The best path taken at the time of call setup is through the Cincinnati switch.

In the second switched-circuit routing scenario, the CO and the IXC switching centers for both the originating city and the terminating city are the same. However, a direct route is taken from New York to Los Angeles. Therefore, the best path taken at the time of call setup is a direct route from New York to Los Angeles.

In the third switched-circuit routing scenario, the CO and the IXC switching centers for both the originating city and the terminating city are still the same. However, the best path taken at the time of call setup is through Cincinnati, Dallas, and Las Vegas.

All three switched-circuit routing scenarios will work. It is fairly obvious that scenario #2 is best because it is the most direct route. It is also the route that uses the least switching equipment. Scenario #3 uses the most equipment and takes the most indirect route.

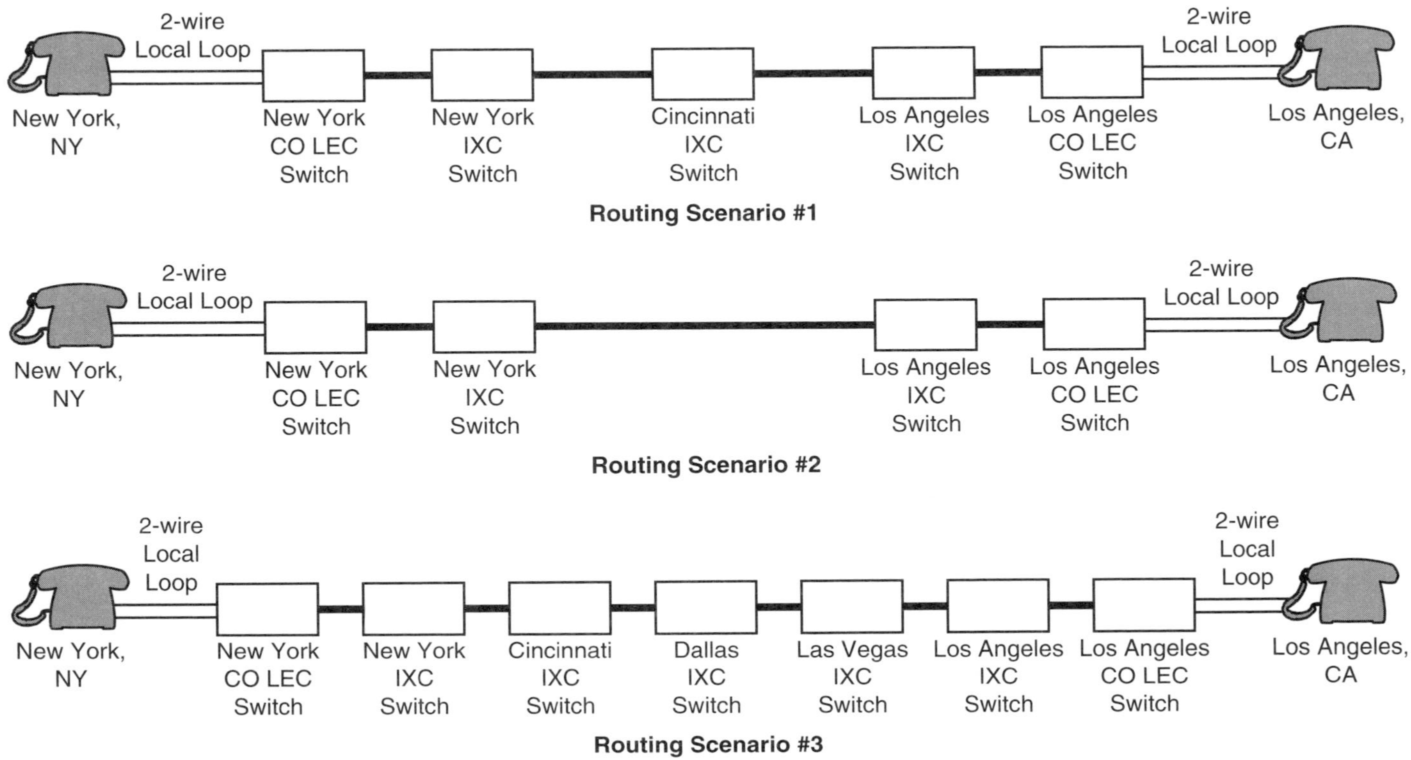

Figure 2.11
Different routing scenarios at the time of call setup.

It is important to realize that all of the above scenarios will work and that each of these scenarios is transparent to the telephone user. Circuit-switched routing paths are unimportant to the telephone user. They are important as far as the network is concerned. Let's examine these important concepts further.

Since a switched circuit is established at the time of call setup, the telephone goes off-hook and the call is established. For a short telephone conversation, this may not be much of a network problem. For longer telephone conversations (more than a few minutes) or Internet connections via a modem, the length of time spent on a call can have some network impact.

Let's say that scenario #3 is established at call setup. The switched service network has determined that this is the best routing or circuit path at the time of call setup. Two minutes later, however, a direct routing path becomes available between New York and Los Angeles. The direct path cannot be used because the circuit is established at call setup and remains with the same configuration for the life of the call.

This process is inefficient as far as the network is concerned. Circuit-switched services are not dynamic. However, the network and circuit availability changes constantly, and switched-circuit services must remain for the life of the switched call.

All of this discussion surrounding the inefficiencies of the telephone network may be true, but there are many positive aspects to a circuit-switched network. The voice quality or quality of service (QOS) is very good. The circuit-switched network has become the standard for voice quality in the industry. Features such as caller-ID, call forwarding, and call waiting have been integrated into circuit-switched telephone services. And, the availability and reliability of the switched-services telephone network are very high.

Until recently, there wasn't any thought about replacing the circuit-switched telephone network with an internet protocol network or IP network, hence the name Voice over IP (VOIP). More about VOIP services when we discuss telecommunications futures in Chapter 10.

Blocked Calls

We already know that the switched-circuit telephone network provides good service. However, in normal circumstances, only a small percentage of telephones are in use at any given time. Therefore, switches are designed to handle a normal amount of telephone traffic on a given day. Any telephone calls the telephone network cannot handle are called "blocked" calls.

In the United States, telephone blocked calls seldom occur. The signal provided by the telephone company is usually a "fast busy" or a prerecorded operator's voice.

Most of the blocking occurs on a holiday, especially on Mother's Day, when telephone calls are at their peak. Generally, measurements are taken on this day and other peak days to determine if more trunks, equipment, or facilities are needed.

If a disaster, such as an earthquake, hurricane, or flood, strikes, the telephone company may not be able to handle the situation (because it is not designed to do so). A "fast busy" signal or no service at all may be the case because all available trunks will be used up quickly in the emergency. Moreover, many local loops are on telephone poles where they are susceptible to the elements.

Caller volume may also be measured during the busy hour. The busy hours occur on a weekday between 10 A.M. and 11 A.M. and between 2 P.M. to 3 P.M. To be cost effective and efficient, circuits are designed to handle slightly more than 100% of the busy call volume to minimize blocking.

Telecommunications network designers use Erlang B traffic tables for PSTN voice call modeling. They use these tables to predict the probability that a call will be blocked.

Capacity Planning

Studies concerning blocking and available trunks are called capacity planning. The objective is to minimize blocking and provide the caller with network access. If the network is overdesigned, there will be too much capacity and wasted bandwidth. If the network is underdesigned, blocking will occur. A good percentage of call volume for the telephone network is between 95% and 120%. Any average above or below this range should be considered unacceptable for today's telephone environment.

For example, a friend of mine wanted to place calls to Puerto Rico with a well-known IXC and was almost always getting blocked. My friend eventually gave up and switched his IXC carrier and was able to get through most of the time. This happens quite frequently with international calls. The carriers do not want to lose business. They will usually try to complete calls even by using a competitor's trunks and facilities (and pay the competitor a fee). It is imperative to keep the customer happy and keep the brand name respectable.

Not only should circuits be available most of the time, they should also be reliable. Noisy circuits, crosstalk, and low volume all contribute to customer dissatisfaction. In addition to voice, data should be virtually error free with few (if any) dropouts.

TELEPHONE SIGNALING

In addition to carrying voice, the telephone lines carry signals used to indicate their status. Some of these signals, such as ring-tone and busy-tone, have already been discussed. Let's take a closer look at signaling and determine how the telephone network uses signaling to process calls effectively.

Basic Signaling

Today's telephone systems use electronic switching systems to interface with the local telephone lines and equipment. This equipment can send and receive signals from the telephone network to the telephone user.

When you go off-hook (or pick up the phone), the electronic switch in the central office senses the circuit completion. It then plays a dial-tone sound so that you know the switch and your phone line are currently working and you are able to proceed. The dial-tone sound is simply the combination of 350 Hz and 450 Hz.

You then dial a number from your touch-tone keypad or DTMF dialer. We already know that the numbers dialed are a pair of frequencies for each key. If the number is busy, you hear a signal that is 480 Hz and 620 Hz with a cycle of a half-second on and a half-second off provided by the central office switch. If the number is available, however, you would hear a ring signal, which is a combination of 440 Hz and 480 Hz.

Can you think of any other types of signals used in the telephone network?

We now know that analog signaling is used through the telephone network even though most of the network (especially the backbone) is digital. This signaling information must be effectively transmitted to the network and ultimately to the terminating switching office to complete a call (Figure 2.12).

We next briefly discuss two types of signaling methods used for voice transmission in the telephone network.

In-Band Signaling

The DTMF or pulse dial signaling, transmitted into the telephone network, can be handled within the same channel as the voice information. This method of signaling was used in the telephone network before the 1970s. Simply put, in-band signaling is

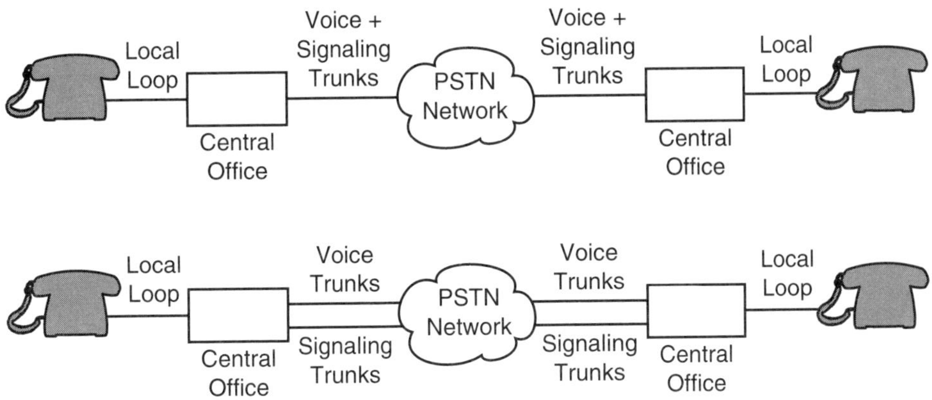

Figure 2.12
In-band and out-of-band signaling.

the sharing of the same circuit or path with the voice signal in the backbone or trunk telephone network. Both the signaling information and the voice information are transmitted together.

In-band analog signaling technology generally furnishes a 2600 Hz tone on each channel between switching centers and the central offices indicating idle status. In-band digital signaling technology may use signaling bits instead of analog tones. Trunk seizures remove the 2600 Hz tone, also known as SF, or signaling frequency, tone.

This type of signaling was used for dedicated or switched services. However, in-band signaling was slow and inefficient. Channels were tied up with signaling transmission data that could be used for voice transmission. Therefore, call set up in an in-band signaling network required additional overhead.

More importantly, security concerns were associated with this type of network. In the 1970s, "blue boxes" were used to simulate in-band signaling frequencies on the network. This resulted in a loss of revenue for AT&T and became a catalyst for migrating the network to out-of-band signaling.

Out-of-Band Signaling

Out-of-band signaling, or common channel signaling, uses a separate packet-switched network to pass call setup and supervisory information. The signaling network determines the best route available to complete the call. Once this route is determined, the voice channels are set up and the call is connected to the telephone end terminal.

With out-of-band signaling, the equipment makes a separate path assurance test before placing the call to ensure end-to-end connectivity. Many of the features associated with this type of signaling would not be possible with in-band signaling. Also, call setup for out-of-band signaling is extremely efficient, and the theft-of-service issues are minimized.

The ITU standard for this type of signaling is Signaling System Number 7 (SS7).

Signaling System Number 7

SS7 is known worldwide as the signaling standard for voice telecommunications using out-of-band signaling techniques. Some of the advantages of SS7 are:

- Call setup, call management, and call teardown
- Wireless service capability
- Handling of addressing and supervision
- Local number portability
- Enhanced calling features (call forwarding, caller ID, three-way calling)
- Toll-free (800/888/877) and toll (900) services
- Support for ISDN services

SS7 messages are exchanged between network elements through 64-kbps channels called signaling links.

The system has three major sections: the service switching point (SSP), the signal transfer point (STP), and the service control point (SCP). SSPs are switches that originate and terminate calls with software algorithms. The SSP can be programmed to use an alternate routing number if the primary number is busy or if the call is unanswered within a specified time. STPs are packet-switching nodes. An STP routes each incoming message to an outgoing signaling link according to the best route information. SCPs are databases containing circuit routing and customer information (Figure 2.13).

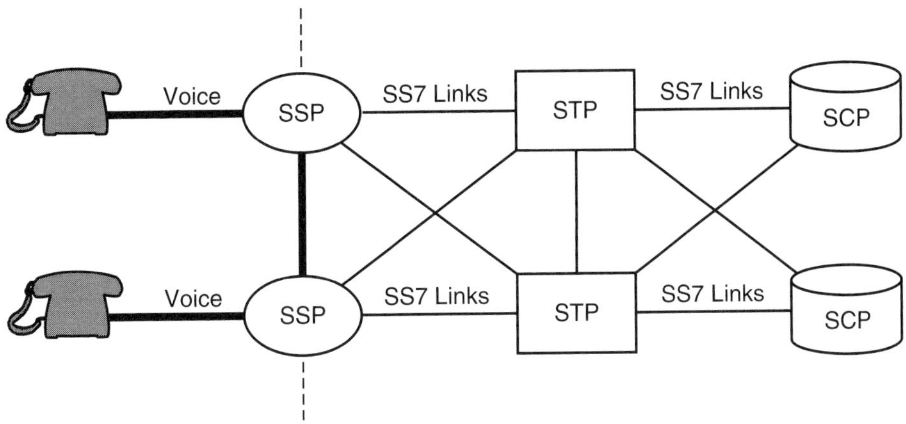

Figure 2.13
SS7 integration.

Call Setup

Now that we understand the basics of the switched services out-of-band signaling, let's see what happens when we use touch-tone dialing to place a call. For example, the number 1-213-555-1234 can be dialed from New York with a final destination in Los Angeles (Table 2.1).

All of this circuit building happens as soon as the first number is dialed on the touch-tone pad. Once the dialed number is completed, the connection is virtually instantaneous.

TELEPHONE NETWORK DATA SERVICES ....

The easiest and least expensive way to provide data services to the home is to use modems on telephone dial-up lines. Modems have evolved rapidly over the last 10 years because of requests for faster speed and larger bandwidth access. The availability of inexpensive personal computers has fueled the drive for better access to the data network.

Table 2.1 NY-to-LA Switched-Services Dialing [1-213-555-1234]

Numbers Dialed	Event
	Originating user goes off-hook and receives a dial tone from the local central office switch through local loops.
1-	Signal to the central office that this number will be a 10-digit long distance number. A circuit to the customer's default long distance carrier can be built at this time.
213-	Signals to the central office that a circuit needs to be built between the originating central office in NY and the customer's default long distance carrier's office in LA.
555-	The circuit can now be extended from the customer's default long distance carrier's office in LA to the LA central office switch associated with the 555 exchange.
1234	The circuit can now be terminated through local loops into the LA phone. An on-hook or off-hook condition will determine whether the call can be completed or a busy signal will result.

Today, standard dial-up modems can achieve speeds up to 56 kbps on a standard telephone analog line. There are also ISDN modems. These modems operate as a digital circuit and can achieve speeds from 64 kbps and up.

The desire to communicate with digital end terminals (such as computers) was realized back in the 1960s with the acoustic-coupler Carterphone. When personal computers became popular in the mid-1980s, modems became necessary devices to enable telecommunications with the outside world. Let's analyze modems in more detail.

Modems

The name modem is short for *mo*dulator/*dem*odulator. AT&T established early modem standards. In the 1980s, standards were set by the ITU. Modems are designed to interface with analog lines or the PSTN. Modems offer the following benefits:

- Voltage correction
- Error detection and correction
- Distortion reduction
- Data compression

- Timing source

Modems have become faster and smarter in a very short period of time. The price of modems has also dropped considerably. Several years ago, 9.6 kbps modems were considered standard. Today, 9.6 kbps modems are considered obsolete.

The function of a modem is to convert a digital signal to an analog signal (D-to-A), and vice versa (A-to-D). Modems enable the transmission of digital information over dial-up analog telephone lines. An example is a 300 bps modem showing which frequencies represent digital signals (Figure 2.14).

Dial-up modems for a personal computer can be internal (through an expansion slot) or external (through a serial connector). PCMCIA modem cards (for laptop computers) are also available.

Modems are designed for unattended operation, emulating a telephone end-terminal device. Modem features include the following:

- Dial-tone recognition
- Automatic dialing
- Automatic answering
- Automatic redialing
- Call termination

When an originating modem (for example, your home) connects to a terminating modem (your Internet service provider), signals are exchanged (modem handshake) to agree on the speed and error correction capability that will be used for

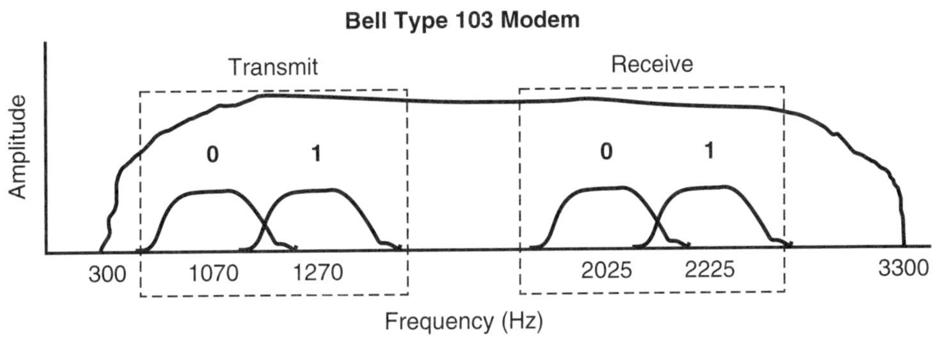

Figure 2.14
Modem frequency conversion through local loops.

Table 2.2 Modem Speed Standards and Comparisons

Standard	Modem Type	Speed	Comments
V.21	dial-up	300 bps	Also used by group 3 fax
V.22bis	dial-up	2400 bps	Compatible with V.22
V.23	dial-up	1200 bps	Has a 75 bps reverse channel
V.27ter	dial-up	2400 bps	Also used by group 3 fax
V.32	dial-up	9600 bps	Also used by group 3 fax
V.32bis	dial-up	14.4 kbps	Has rapid rate negotiation
V.32ter	dial-up	19.2 kbps	AT&T specification
V.34	dial-up	33.6 kbps	Initial speed was 28.8 kbps
V.90	dial-up	56 kbps[a]	Asymmetrical

a. 33.6 kbps upstream
56 kbps downstream

transmission (Table 2.2). If noise and faded signals exist on the line, the speed of a modem may be lowered.

The industry standard is 56 kbps, or V.90, modems. These modems are asymmetrical in speed. In other words, the downstream data (data from the network) is higher than the upstream data (data to the network). It is assumed that the average user receives more data and graphics than he or she sends.

The condition of the local loop, the distance of the local loop from the central office, and the central office power requirements determine whether 56 kbps throughput can be achieved. Throughput is the error-free information transmitted over a network. Generally, 56-kbps modems can achieve a high speed of 53 kbps, with a low of 33.6 kbps used as a fallback speed.

It is important to understand that modems haven't changed much from their original design. We still have an analog 3 kHz telephone line to deal with, and we still must perform an analog-to-digital conversion in each direction. The major change in modems is due to compression algorithms. Without compression, modems would still be operating at 2400 bps. More about digital signals and compression in Chapter 3.

BRI-ISDN

ISDN, or integrated services digital network, has been available since the advent of SS7 in the 1970s. The service provides high speeds to the home for data transmission. In addition to data, digital voice can also be transmitted when digital telephones are used. ISDN standards did not exist for several years and central office digital switches are necessary to allow this service to operate.

BRI (basic rate interface) ISDN is available in areas where some of the newer technologies (DSL, cable modem, etc.) are not. ISDN uses existing 2-wire local loops for transmission, making the service readily available.

The key incentive for ISDN is faster data transfer handling and higher speeds than are provided by traditional analog dial-up methods. For the small office user, this can be a welcome service, especially for bandwidth-hungry applications (such as the Internet). ISDN lines can also provide video-conferencing and backup services.

As workers begin to spend more time telecommuting from home, ISDN can be an economical method of obtaining reasonable speeds at a cost-effective rate. ISDN at one time was billed at a per-minute rate. Today, several LECs offer standard monthly flat rates at a reasonable cost.

BRI-ISDN offers two 64-Kbit digital channels or B-channels (bearer channels) that can be used for voice or data. An additional D-channel for signaling is also used. A 64-Kbit channel can be used for uncompressed digital voice. Therefore, a digital telephone can be used for one channel, while 64-Kbit digital data is used for the other channel.

The 2 B-channels can be bonded together. That is, they can be used as a 128-Kbit digital data channel to provide faster access.

The D-channel remains as a signaling channel operating at 16 Kbits. Note that out-of-band signaling is used. The signaling channel always remains separate from the bearer channels (Figure 2.15).

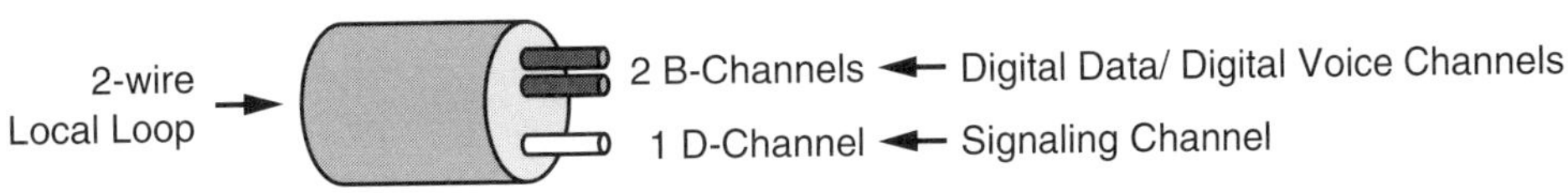

Figure 2.15
BRI-ISDN.

CONCLUSIONS

This chapter has summarized the events and technologies associated with plain old telephone services. While many new technologies remain on the horizon, standard telephone service has remained remarkably stable, with extremely high reliability and availability rates.

The big problem with POTS is the 3 kHz bandwidth limitation for an analog voice grade channel. We've reached the limitations with modem speeds at 56 kbps. Therefore, we must look to new technologies and be prepared to re-engineer the network to provide faster data services.

The networking costs of providing new high bandwidth services are extremely high. In many situations, we must work with technologies such as 56-kbps modems and BRI-ISDN, until new services become available.

The next chapter takes a closer look at digital services. The basic building blocks of digital services begin with the T-1 rate. We'll take a look at how these services operate and what systems are available.

3 Telecommunications in the Digital Age

Thus far, Chapter 1 was concerned with basic telecommunications concepts and principles, and Chapter 2 provided us with the knowledge of basic telephone services. We now need to concern ourselves with today's voice, video, and data applications.

Before we can proceed to high bandwidth applications, there are still more telecommunications topics we need to cover. We need to understand digital sampling and digital data transmission. We also need to understand connections between networks and how networks transmit to and receive from one another. We also know that telecommunications systems can differ depending upon the type of media used, the bandwidth, protocols, and whether the signals processed are analog or digital.

Digital is the language of computers and analog is the language of humans and the world around us (music, sound, etc.). It is important to understand the differences that both analog and digital have to offer. Therefore, we must recognize that analog-to-digital conversion (and vice versa) will always be necessary.

Digital technology has replaced analog technology in all aspects of transmission systems except the PSTN local loop. Digital transmission is a two-state system (1's and 0's). It is easier to troubleshoot and maintain than an analog transmission system. Digital transmission also has the advantage of a direct interface into modern switching systems without the use of analog frequency conversion. Digital systems do not propagate noise, and the quality is superior to analog systems.

This chapter is divided into five sections. First, we examine digital-to-analog conversion, T-1 and T-3 transmission rates, and TDM services. In the second section, we reexamine telecommunications connections: twisted pair, coaxial, and fiber cable. In the third section, we explore transmission systems, including four cabling topologies: bus, star, ring, and mesh. We also look at backup networks: backbone and redundancy, circuit systems, PRI-ISDN, and baseband and broadband systems. In the fourth section, we discuss transmission methods, such as synchronous/asynchronous transmission and simplex/half-duplex transmission. The fifth section covers voice, MP3, and MPEG1, 2, and 3 compression requirements for digital transmission.

A 0 and a 1 and a…

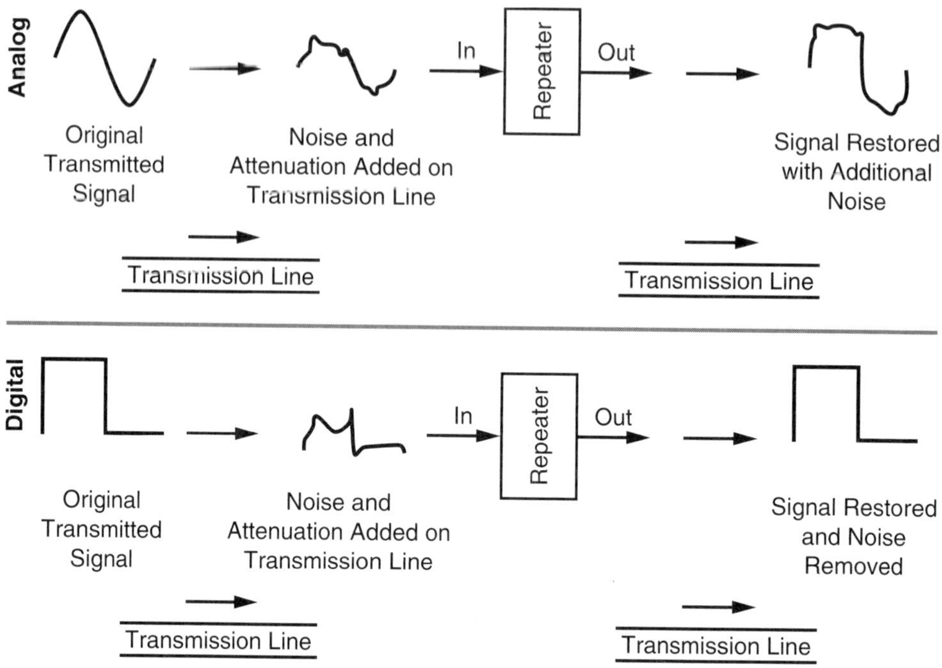

Figure 3.1
Analog vs. digital signal regeneration.

DIGITAL TRANSMISSION SYSTEMS ....

If a circuit is designed correctly, a digitally transmitted signal can arrive at the receiving terminal with unsurpassed quality. This is one of the reasons why an international phone call can sound as clear as the neighbor's call from next door. Analog transmissions contain noise, which is additive. The longer the transmission wire, the more noise added to the circuit. Analog repeaters cannot resolve this problem. Actually, analog repeaters add to the problem by boosting both the signal and the noise simultaneously.

It should be noted that digital repeaters can be spaced farther apart than analog repeaters. Also, a digital repeater actually functions as a regenerator. The digital repeater/regenerator restores the weakened square wave signal back to the original. Of course, the circuit must be properly engineered to regenerate the signal to its original state (Figure 3.1).

We've already mentioned that digital transmission is a two-state system where a high can represent a 1 and a low can represent a 0. Therefore, a 1 then a 0 would appear as a square wave.

TIMETABLE OF HISTORIC DIGITAL TRANSMISSION EVENTS

1938	Alex Reeves (ITT-UK) patents pulse code modulation method for converting analog to digital.
1948	Claude Shannon outlines a series of mathematical formulas to reduce telecommunications transmission to binary code (1's and 0's).
1960	Following the invention of the transistor, "digitized" voice is developed.
1963	AT&T uses digital backbone T-1 rate transmission for the PSTN.
1983	Consumer/business applications of digital T-1 rate are first applied.

Digital Binary Data

Let's discuss the differences between the decimal system, or base-10 system, and the binary system, or base-2 system.

In the decimal system, counting from right to left, we normally start at the first column counting 0, 1, 2, 3, 4, 5, 6, 7, 8, 9. After 9, we add one digit to the second column and restart at 0 (in the first column) making the number 10 or ten. In the binary system, we start at the first column counting 0, 1. After 1, we add one digit to the second column and restart at 0 (in the first column) making the number 10 or two.

Therefore, the decimal system has columns of 1, 10, 100, 1000, 10,000, etc., and the binary system has columns of 1, 2, 4, 8, 16, 32, 64, etc. If you multiply each column by 2 you will obtain a result for the next column (Table 3.1).

Table 3.1 Decimal Binary Comparisons

Decimal Base 10				Binary Base 2						
1000's	100's	10's	1's	64's	32's	16's	8's	4's	2's	1's
0	0	7	0	1	0	0	0	1	1	0
seventy (70)				seventy (70)						

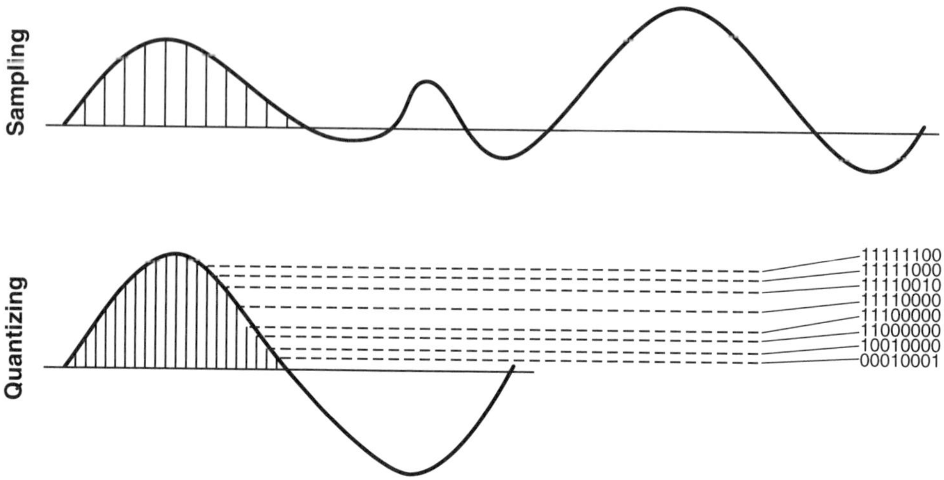

Figure 3.2
Voice sampling and quantizing.

Sampling with Nyquist's Theorem

It is possible to take incremental samples of a waveform and store the result as a digital value (1's and 0's). The rate of samples taken is the key. If we take too many samples, we are overdesigning an analog-to-digital converter. If we take too few samples, we risk the possibility of losing critical information (see aliasing below). We can sample analog frequencies into a digital format. We can then transmit them in an attempt to reconstruct the signal at the receiving end or store them in a computer for future use (Figure 3.2). The more samples taken, the more information we need to transmit or store.

Just how much information should we sample? Nyquist's theorem states that an analog signal must be sampled at a rate twice the highest frequency (Table 3.2).

Table 3.2 Sampling: Nyquist's Theorem

Type	Frequency Range	Sampling Rate
Analog human voice	100 Hz – 10 kHz	20K times/sec
Analog recorded music	20 Hz – 20 kHz	~44K times/sec
Analog telecom voice channel	20 Hz – 3.3 kHz	~8K times/sec

If you look at audio specifications for a CD player, you'll notice sampling rates of 44.1 kHz, or 44,100 samples per second. That's a lot of sampling to accurately reproduce analog music. With today's miniaturized integrated circuits we can have electronic switching devices that will sample at this rate. Therefore, the CD sampling rate of 44.1 kHz will capture frequencies up to 22 kHz (a little past the 20 kHz specification to ensure no loss of high frequencies).

What happens if the sampling rates are too low? We encounter a problem called aliasing. Aliasing seems as if you sampled a low frequency that probably wasn't part of the original signal. Alias frequencies make a lot of noise and distort the original signal. Oh yes, a low sample rate also removes the highest frequencies and mixes these high frequencies with more noise and even more distortion.

Quantizing and Companding

The amplitude of the sampling pulse is encoded into an 8-bit sample by a process called quantizing (Figure 3.2). An 8-bit sample provides 2^8 or 256 steps. Each step corresponds to a particular amplitude of a sample. Since the output of all the samples (or steps) is designed to reproduce an analog waveform, the result is a waveform with a series of samples or steps and not the original analog waveform that is intended (Figure 3.2). When we listen to this conversion from an analog waveform to samples and steps, we hear an effect known as quantizing noise. Quantizing noise is most audible at lower frequencies and can be overcome by a process known as *companding*. Companding is a form of compression and decompression used to smooth a signal.

T-1 Rates

T-1 rates have become a standard in the industry. These rates were originally developed by AT&T in the 1960s to "digitize" their backbone network. The design calls for 24 telecom voice circuits over one digital path. T-1 rates were not available to the consumer or business sector until the mid 1980s. Today, T-1 (also known as DS-1) circuits can transmit a combination of voice, video, and data.

Let's step back for a minute. We already know that a telecommunications voice channel is 3 kHz. Using Nyquist's theorem, we can sample voice at a rate of 8000 times a second. Therefore, if we multiply 8 bits a sample by 8000 samples a second, we have a rate of 64 kbps. This is our standard digital channel rate or DS-0 level. The prefix DS stands for "digital signal." It should be noted that the telecommunications carrier may subtract 8 kbps for signaling and maintenance. This leaves 56 kbps for the customer. If the full 64 kbps is available, the circuit then contains "clear channel signaling" and the 8 kbps are not lost.

To derive our full T-1 rate we would multiply 64 kbps times the standard 24 channels and add 8000 framing bps. This provides us with the 1.544 Mbps full T-1 rate (Table 3.3).

Table 3.3 T-1 Rates

8,000	----------samples per second	
× 8	----------bits per sample	
64,000	----------bits per channel	——► DS-0 Level
×24	----------number of channels	
1,536,000	----------bits per second	
+8,000	----------framing bits per second	
1,544,000	------------bits per second total	——► T-1 or DS-1 Level

T-1 generally runs well over twisted pair copper. Two wires are used for the transmit operation and two wires are for the receive operation.

Unfortunately, T-1 rates used in the United States, Canada, and Japan are not compatible with E-1 rates used in the "rest of world." E-1 rates use 2.048 Mbps with 30 channels digitized.

Fractional T-1 Rates

Customers who require less than T-1 rates will often look for Fractional T-1 rates to save money. Fractional T-1 is sold in multiples of DS-0 channels (64 kbps each). For example, five DS-0's or 64 kbps × 5 = 320 kbps. Customers with multiple locations can also use a drop-and-insert device. For example, three DS-0's could be used for communications between the main office and a branch location. Two other DS-0's could be "dropped" into another branch location from the main office (Figure 3.3).

The more DS-0's you have, the more the cost approaches a full T-1 rate. Therefore, a good rule of thumb is a Fractional T-1 containing five DS-0's. If more than five DS-0's are needed, a full T-1 rate would provide more channels at the same cost.

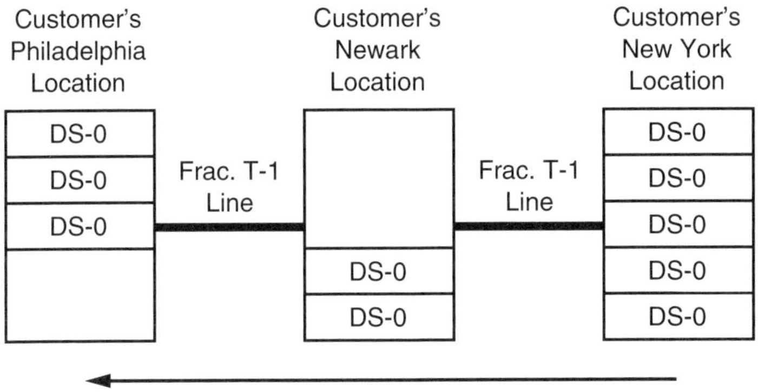

Figure 3.3
Fractional T-1 drop-and-insert.

Time Division Multiplexing

We've already covered frequency division multiplexing (FDM). Time division multiplexing (TDM) works on much of the same principles as FDM, only we divide segments of time instead of segments of frequency. In TDM, each station is assigned a time slot (TS). TS #1 would sample voice call #1 for several milliseconds and then move on to TS #2, where the sampling would continue etc. The TS multiplexer would simultaneously gather the data from all of the time slots and send out the information as one bitstream (Figure 3.4).

An enhancement to TDM is the statistical time division multiplexer (STDM). This equipment is sometimes referred to as a statmux. An STDM attempts to make the TDM more efficient by assigning time slots only when they are in use. For example, if a station has no data to send, the time slot will be reassigned to someone else. STDMs improve circuit utilization by keeping idle transmission to a minimum. STDMs are more expensive than TDMs and must be monitored to prevent TS overload.

STDM can be used for voice or data. Human speech or voice may be present during 80% of a typical transmission (as an average) with silence during the other 20% of the transmission. A voice TS could be reassigned during the 20% intervals of silence. Data can also be sent through an STDM. Much of today's data is "bursty" because of its "on again, off again" nature. Generally, STDM works very well with any type of data.

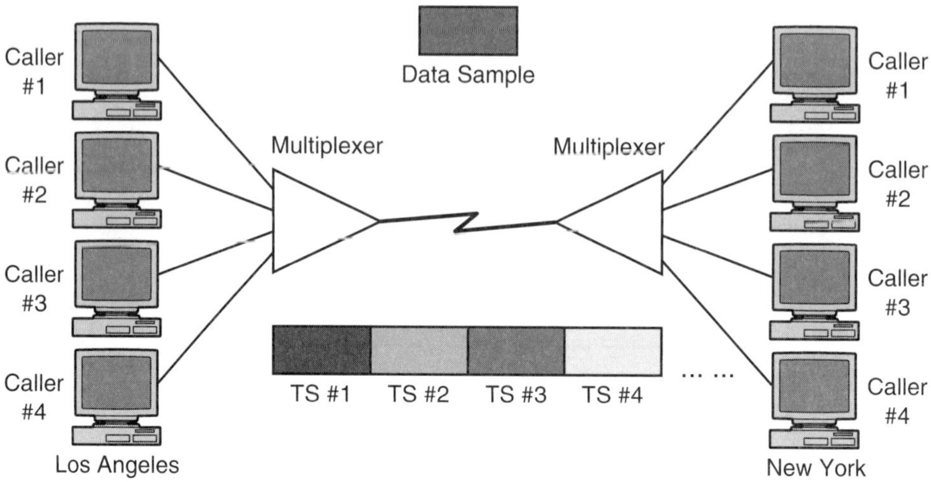

Figure 3.4
Time division multiplexing.

T-1 Channel Bank

We mentioned before that T-1 rates can provide telecommunications flexibility for the consumer. A device is needed to convert the digitized information on the T-1 transmission line to analog. This device is called a channel bank. A channel bank can provide the flexibility needed to handle voice, video, and data applications. For example, you can bond five DS-0's together and transmit an uncompressed video channel. Nineteen other DS-0's would still be available for voice or data.

Channel bank manufacturers provide many options for these devices. Many signaling options are available with plug-in modules for channel units. Some of these channel units have provisions for telephone ring-down and automatic gain control (increasing decibel levels).

Because of the T-1 channel bank device, the 1.544-Mbps T-1 rate is now customizable; it is still a very powerful, affordable choice for businesses and small offices relying on telecommunications.

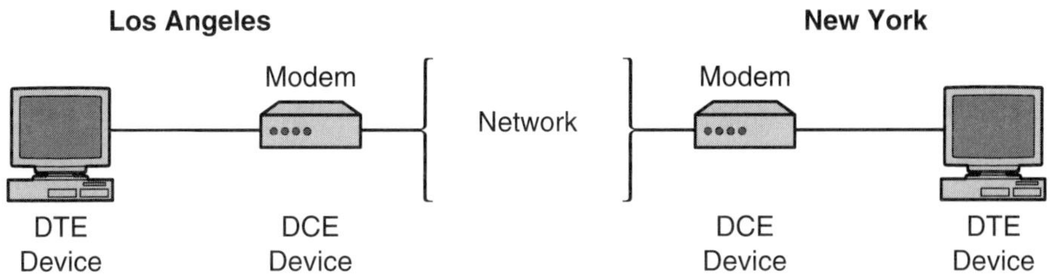

Figure 3.5
DTE and DCE.

T-3 Rates

T-3 is the next step beyond the T-1 rate. T-3 consists of 28 T-1's, which is the equivalent of 45.736 Mbps or 672 DS-0's. T-3 rate can be very cost effective in comparison to the cost required for 28 T-1's. T-3 rates can be desirable for large businesses, college campuses, or telecommunications backbone networks.

Note that T-3 rates do not use twisted pair copper. Generally, a coaxial cable is used for transmission. Of course, fiber will handle these types of transmissions also.

DTE and DCE

The user's device in a data network is usually referred to as data terminal equipment (DTE). A DTE can be end-terminal equipment such as a computer, or it can be a simple "dumb" terminal. The equipment used to interface between an end-user's terminal and the network is called data circuit equipment (DCE). Examples of DCE are modems or multiplexers (Figure 3.5).

DSU and CSU

A data service unit/channel service unit (DSU/CSU) connects a DTE to a telecommunications circuit by providing the correct bitstream required for digital transmission. In addition to a digital conversion, the DSU/CSU can also provide a loopback function. A loopback function provides a loopback point for a telecommunications carrier to test a circuit without physically going to a customer's office to perform a test.

DSU/CSU functionality need not always be a separate device but can be included in another device such as a router.

TELECOMMUNICATIONS CONNECTIONS ....

Let's now look at the cables, network configurations, and connectors needed to access a telecommunications network.

Standard Twisted Pair Connections

The most common telecommunications device in use today is the RJ-11 connector. This plastic device is used for a connection of a user device or phone to the PSTN and contains two conductors (two-wire circuit) connected to twisted pair copper.

The RJ-14 connector is the same as RJ-11 except that RJ-14 connectors contain four conductors and are generally used for 2-line phones (two 2-wire circuits).

Data circuits generally use a type of connector similar to the RJ-11/RJ-14; however, the number of conductors this time is eight. These connectors are called RJ-45. It is possible for data connections to use a balun, which is a device that converts coaxial wiring to standard twisted pair wiring and vice versa. RJ-45 connectors can be connected to network interface cards (NIC), which interface with a computer. We'll revisit data protocols—Ethernet and token-ring—toward the end of this chapter.

Wires in a connector can be organized in various configurations (Table 3.4). Be sure that you are using the correct wiring configuration if you are creating your own cables and connectors.

Both types of connectors are made of plastic with metal conductors or "pins." To use them, you first strip off some of the insulation, exposing the bare copper wire. Next, after lining-up the wires correctly, you use a crimping tool to squeeze the "pins" in such a way that they do not contact the bare end of the wire (Figure 3.6).

In addition to the above, there are versions of connectors for stranded conductor and solid-conductor cables. Stranded conductor cables usually contain tiny strands of copper wound together with thread. This type of conductor is difficult to modify for connection. Cables containing this type of wire are delivered from the factory with connectors attached directly.

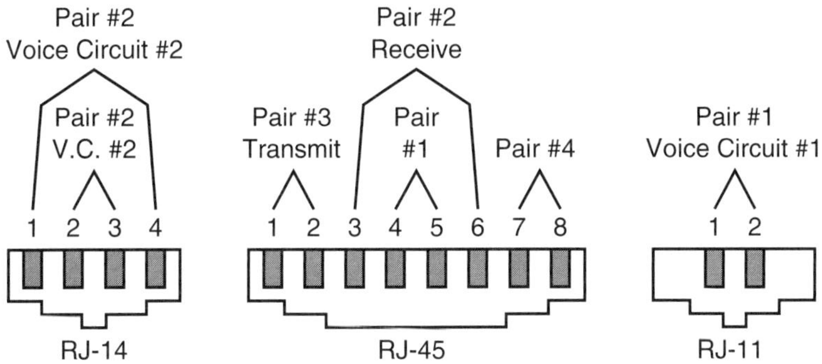

Figure 3.6
Twisted pair connectors.

Table 3.4 RJ-45 Typical Color Coding Arrangements

	Standard 568A	**Standard 568B**	**Crossover**
1	White/green	White/orange	White/green
2	Green	Orange	Green
3	White/orange	White/green	White/orange
4	Blue	Blue	Blue
5	White/blue	White/blue	White/blue
6	Orange	Green	Orange
7	White/brown	White/brown	White/brown
8	Brown	Brown	Brown

Most of the copper cable installed for both voice and data is twisted pair. In local data networks, wires can range from a minimum of two to a maximum of eight in a twisted pair cable. Therefore, we should understand that problems not visible to the naked eye can occur. Any mishandling of copper cable installation can cause problems. A good copper twisted pair wire tester (or multimeter) can detect copper wiring faults as listed below:

- Open pairs—An open condition can occur when one or more wires in a cable are not connected to the pins at the opposite end.
- Shorted pairs—A short condition can occur when wires in a single pair are connected together within a cable.

- Crossed pairs—A crossed-pair condition can occur when a connector does not have the wires installed correctly and the wires of one color path are connected to the pins of a different color pair at the opposite end.
- Reversed pairs—A reversed-pair condition can occur when two wires of a single pair are connected to the opposite pins of the same wire pair.
- Short between pairs—A short-between-pairs condition can occur when one wire of a pair is incorrectly connected to one wire of another pair with a cable.
- Split pairs—A split-pair condition can occur when one wire on each of two pairs is reversed at both ends of the run.

Note that twisted pair (TP) can be described as unshielded twisted pair (UTP) or shielded twisted pair (STP). UTP is standard with no additional wire braid or foil surrounding the wires in a cable. STP can be used where environmental interference can have an effect on voice or data transmission. Areas such as power cable runs, fluorescent lighting ballasts, video monitors, and automotive engines are good examples of interference-causing devices.

Standard Coaxial Cable Connections

As we discussed in Chapter 1, several types of coaxial cable are used in today's telecommunications applications. The process for making coaxial cable connectable is the same for both a BNC connector and an F connector. For local data wiring, a BNC connector is used. This connector has a knurled grip with a quarter turn locking slot. Video applications use a screw connector or F connector.

There are crimping tools for both F connectors and BNC connectors. There are also twist-on or threaded coaxial connectors. Either connector requires that the cable be stripped to a specified length and then inserted into the connector accordingly (Figure 3.7).

Most of the coaxial cables used today are for video applications. Since there are only two wires in a coaxial cable, the cable is easy to troubleshoot. Conditions do exist, however, that can cause a coaxial cable run to fail. We have already covered these conditions in the twisted pair configuration, but they are listed below for convenience:

- Open pairs—An open condition can occur when the central conductor, braided shield, or both wires in a cable are not connected to the pins at the opposite end.

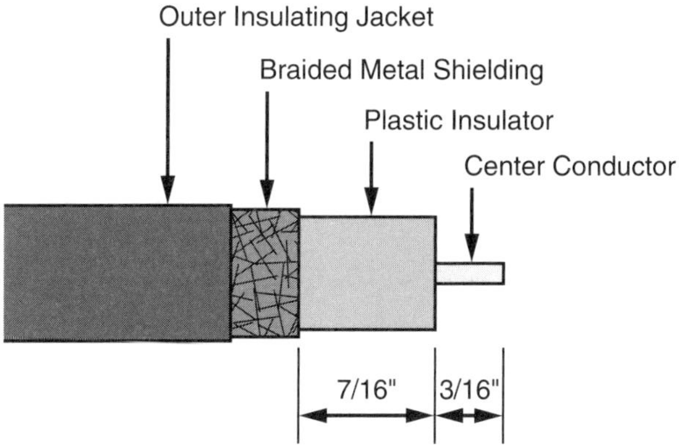

Figure 3.7
Coaxial cable splicing.

- Shorted pairs—A short condition can occur when the center conductor and braided shield are connected together within a cable.

Note that coaxial cable is always shielded. The cable is well suited to repel outside interference, which can have a disastrous effect on signal transmission (see twisted pair section above).

Standard Fiber Connections

Although fiber connectors and cables are not yet standard for the home environment, the perception is that there will be a standard within the next few years. Therefore, it is appropriate to discuss fiber cable and connectors at this time.

Fiber cables have many different types of connectors, and there are several different splicing methods. Unlike copper connections, fiber has a light, which must be aligned properly. Just stripping the fiber cable and attaching a connector isn't enough; a fiber connection must be precise.

Fiber is the most difficult of the three basic types of closed media to splice and make connectable. During the mid-1980s, when fiber first became mass produced, a single fiber splice could take half an hour or more. Today, fiber splicing is done within a few minutes and is becoming a routine operation.

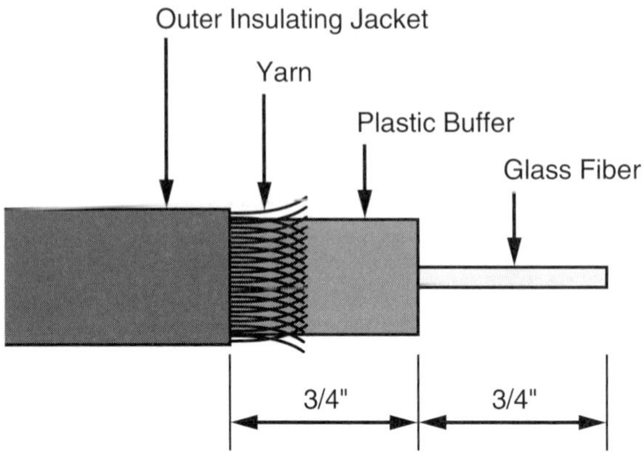

Figure 3.8
Fiber splicing.

There are two types of fiber connectors: epoxy and epoxyless.

Epoxy connectors can already have an adhesive preloaded into each connector and can be activated with high temperature or UV light. Most of the fiber connectors in use today have an epoxy connector. Unfortunately, this type of connector requires a drying time of approximately 15 minutes. Hence, the epoxyless connector has been developed. It uses a friction crimping method to hold the fiber in place on a connector.

It is important to understand that fiber contains glass and glass can break or shatter, causing injury. Also, avoid twisting and turning the cable when splicing and avoid any harsh movements because fiber breaks easily. Generally speaking, thoroughly understand the manufacturer's directions before proceeding with any type of fiber splice.

Stripping fiber cable (Figure 3.8) is similar to stripping coaxial cable.

After the epoxy has cooled or the fiber has been crimped, the excess glass fiber end must be cut. A special fiber tool known as a scribe is used to trim the glass fiber end and provide a clean edge. Finally, for a proper connection, the glass fiber end must be polished to a flat level surface. Polishing should last for about 15 seconds. Overpolishing can cause gaps in the connection, making the cable useless.

The smoothness and leveling of the glass fiber end is extremely important. Without proper contact, a loss of light will result and the cable will not function properly.

Today, most of the glass fiber cable is installed for high-speed networks and is not intended for the home office or end user. This situation may change within a few years as fiber costs continue to drop and bandwidth demands continue to increase. Fiber is easily breakable, and while this type of cable is not subject to outside interference, an open condition can certainly occur. Some of the causes of open conditions are as follows:

- Excessive gap between fiber and connectors
- Impurities within the glass fiber core
- Excessive cable bending
- Excessive cable stretching
- Improper splicing

The main factor to watch in fiber optic cabling is attenuation. We've already learned that attenuation is a decrease in power. However, in a fiber cable, we are speaking of a decrease in a light signal, which will adversely affect transmission speed.

PHYSICAL TOPOLOGY

Since we now have some understanding of cabling and associated connectors, we need to understand physical topology. The physical topology of a network is a roadmap describing cable layout and end-terminal equipment. Designing network topology is critical. A poorly designed network in a home or small office can be costly.

Cabling Topologies

Depending upon the situation and the type of data to be transmitted, each topology has its advantages and disadvantages. Let's discuss the four types of cabling topologies: bus, star, ring, and mesh.

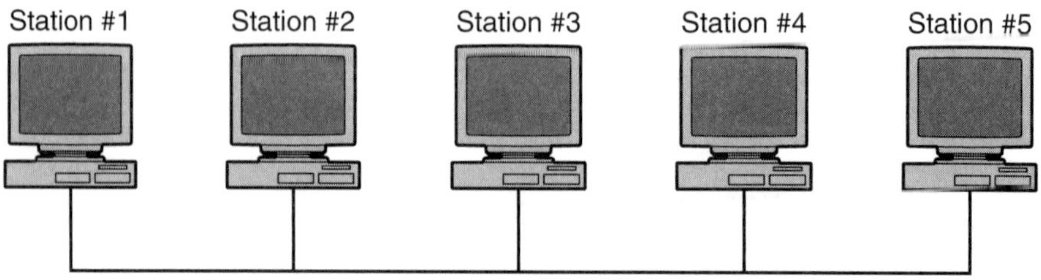

Figure 3.9
Bus topology.

Bus Topology

In a bus topology, all of the end-terminal equipment are connected in series. Drop cables, connected to each end-terminal device, are then connected to a main cable (Figure 3.9).

Advantages of bus topology are as follows:

- Simple installation
- Minimal amount of cabling
- Inexpensive

Disadvantages of bus topology are as follows:

- Difficult to make changes without major disruptions
- Difficult to troubleshoot without major disruptions
- Fault intolerant—a single fault can split a network in two

Major uses of bus topology are as follows:

- Internal components of a device, such as a mini-computer, use bus architecture to communicate with each other.

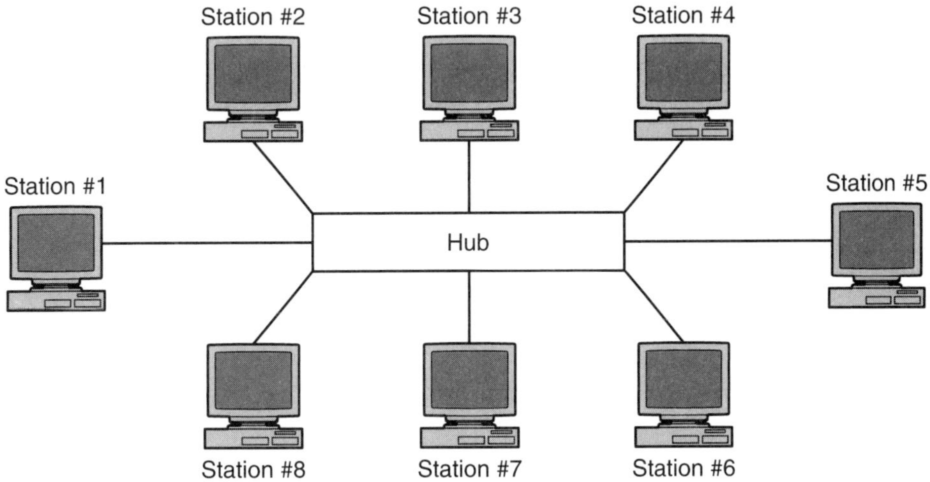

Figure 3.10
Star topology.

Star Topology

In a star topology, the end-terminal equipment is connected to a central hub device. Cables are all connected to a central location (Figure 3.10).

Advantages of star topology are as follows:

- Easy to troubleshoot
- Fault tolerant—single cable failure will disengage one end terminal
- Easily reconfigured

Disadvantages of star topology are as follows:

- Single point of total failure: the hub
- Higher installation cost because of the amount of cabling
- No backup path in case of one end-terminal failure

Major uses of star topology are as follows:

- Local area network

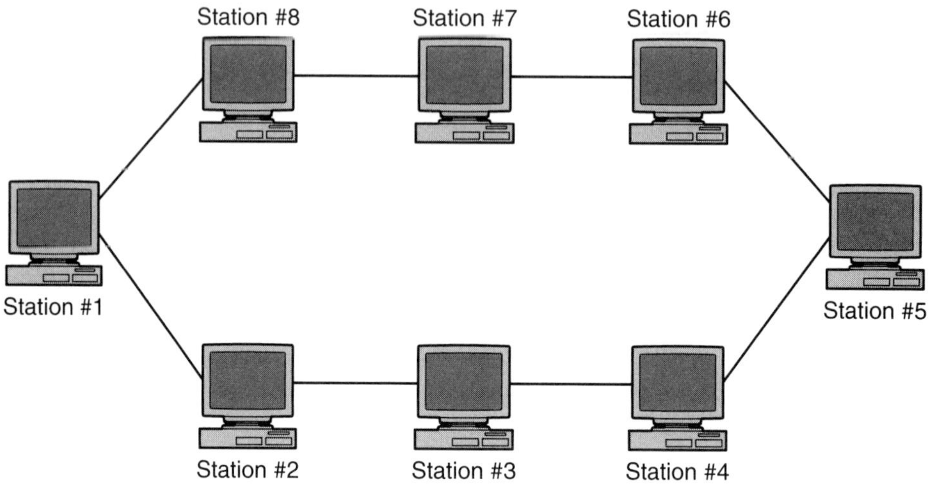

Figure 3.11
Ring topology.

- Central office (acts as a hub) with end-terminal telephone sets
- Home or small office environment

Ring Topology

In a ring topology, the end-terminal equipment is connected to the other end-terminal equipment in a ring fashion. Cables are all connected between end-terminal devices. Generally, a dual-ring topology is used for backup (Figure 3.11).

The advantage of ring topology is:

- Simple installation

Disadvantages of ring topology are as follows:

- Fault intolerant (single ring)—cable failure will disengage entire network
- Expensive to install—multiple cables needed for each end terminal
- Difficult to reconfigure

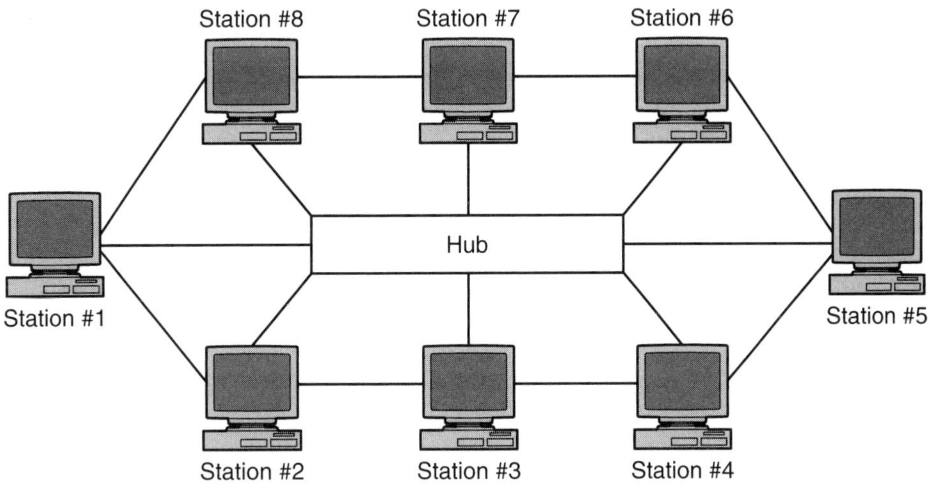

Figure 3.12
Mesh topology.

Major uses of ring topology are as follows:

- Fiber backbone systems (dual ring)
- Local area networks

Mesh Topology

In a mesh topology, all the end-terminal equipment are connected to the other end terminals from different directions in a mesh fashion. Cables are all connected between end-terminal devices. Generally, a mesh topology can used for backup (Figure 3.12).

Advantages of mesh topology are as follows:

- Fault tolerant—any cable failure has built-in backup, so outages are minimized

Disadvantages of mesh topology are as follows:

- Expensive to install—multiple cables needed for each end terminal
- Complex installation

Major uses of mesh topology are as follows:

- Wide area networks
- When fault-tolerance is mandatory

Before choosing the topology needed for your application, *know your data.* Generally, in the home or small office environment, a star network works well and is easy to troubleshoot. Hubs are rarely a problem in this type of network. However, if a fault-tolerant network is your desire and cost is not of utmost concern, you must examine a mesh topology, which contains a backup configuration.

Let's examine backup networks in more detail.

Backbone, Redundancy, and End Segments

A backbone network is the large pipe or trunk to which all segments can connect. The backbone is the main part of a network and generally uses high-speed fiber for transmission. A good analogy of a backbone transmission line is a large water pipe serving a housing development or community.

All backbone networks require a backup, or redundant, path. Redundant paths are needed in case of failure. Failure is unacceptable because hundreds, thousands, or millions of customers could lose service if a fiber cable is cut. It is important to understand that the redundant path must be physically separate from the main path.

Automatic switchover devices can sense a cable break and cut over to a redundant cable with little or no human intervention.

End segments are connected to the backbone network. End segments serve a smaller area than do backbone segments. An example of an end segment is a T-1 line. End segments can use redundancy and should do so for mission-critical data (Figure 3.13).

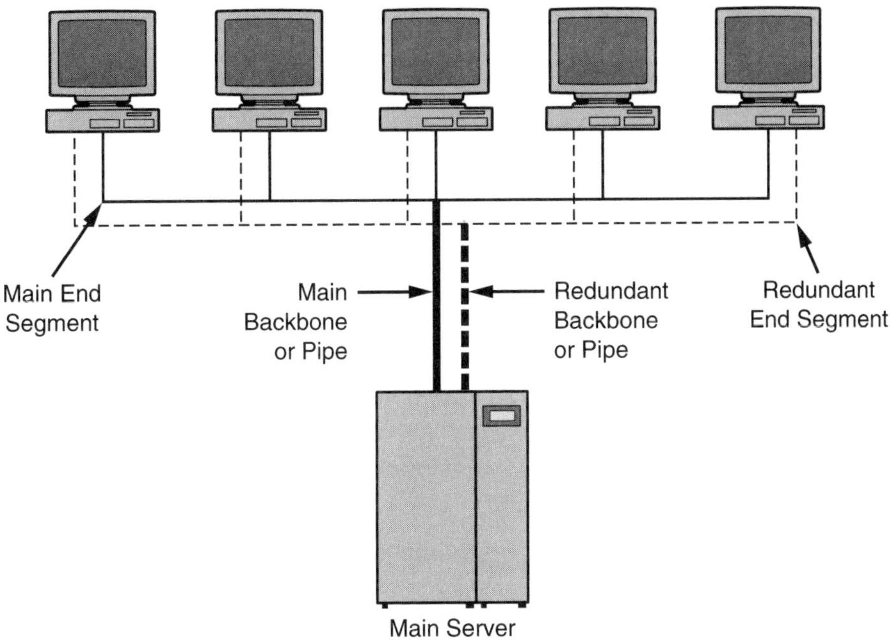

Figure 3.13
Backbone, redundancy, and end segments.

Point-to-Point Dedicated Circuits

A point-to-point circuit is a private circuit with end terminals dedicated to each other. For example, a wired intercom may have an end terminal in the kitchen and an end terminal in the bedroom. It is possible to communicate point-to-point from the kitchen to the bedroom without others listening or adding to the conversation. These terminals are dedicated to each other.

Another example of point-to-point telecommunications is a circuit from the White House to the Kremlin. The U.S. president would speak with Russia's president directly without a network and (we hope) without others eavesdropping or adding to the conversation.

Other examples of point-to-point dedicated circuits are the microwave and infrared transmission systems we reviewed in Chapter 1. Both microwave dishes and infrared lasers are aimed at each other to provide a service from terminal A to terminal Z (Figure 3.14).

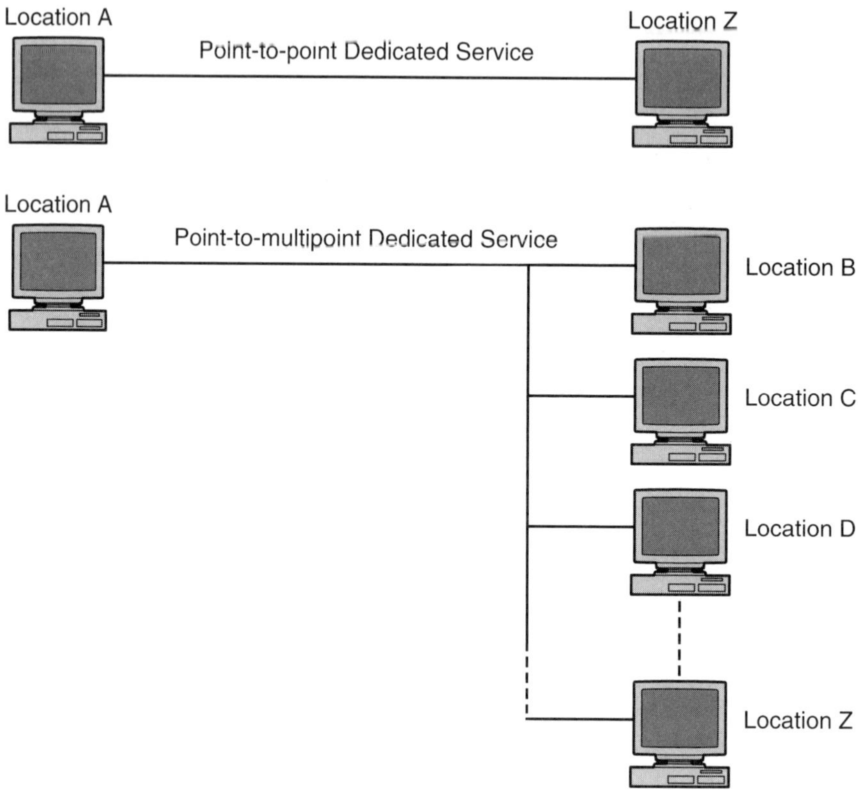

Figure 3.14
Point-to-point and point-to-multipoint configurations.

Costs for point-to-point dedicated services are based upon distance and bandwidth. Therefore, a dedicated 56 kbps circuit from New York to Los Angeles would cost more than a dedicated 56 kbps circuit from New York to Philadelphia.

Humans have "point-to-point" dedicated communications also. This is usually referred to as a one-on-one conversation.

Point-to-Multipoint Circuits

A point-to-multipoint circuit contains a master station, which interfaces with several remote stations in a private network. The remote stations can be programmed to receive all information transmitted from the master. If this is not desirable, each remote station can be assigned a separate address so that each station can receive only its data.

A good example of this type of network is a corporate office (master station) transmission to all remote offices on their private network (Figure 3.14).

Costs for point-to-multipoint dedicated services are also based upon distance and bandwidth (similar to point-to-point pricing). Cost savings will occur because a separate line will not be needed between location A and each remote station.

People have point-to-multipoint dedicated communications also. An example of this is a speech by a corporate president to employees in an audience or a "one to many" conversation.

PRI-ISDN

PRI-ISDN supports T-1 bandwidth at 1.544 Mbps and comprises 23 bearer or main channels and one signaling channel. PRI-ISDN services can be used as a backup switched service. For example, a main T-1 dedicated line may require redundancy, however, the customer may not want to have the expense of a dedicated redundant line that is virtually unused.

Since PRI-ISDN is a switched service, you pay only for what you use. A dial-up number is provided for each end terminal. Upon main circuit disconnect, the customer would dial up the other end terminal and make the connection. Therefore, this service is ideally suited for T-1 backup.

Baseband/Broadband Networks

Today, there is much talk about broadband and baseband transmission systems. The basic difference between baseband and broadband is the transmission method.

Baseband transmissions are unmodulated. A good example of this is a standard fiber transmission. Baseband fiber transmission uses pulses of light to send digital information (1's and 0's). The signal is driven into the transmission medium without any modulation. Baseband is simpler and less expensive than broadband transmission systems.

Broadband transmissions send modulated analog signals (or frequencies) through guided or unguided media. A good example of broadband transmission is cable TV. Each TV channel is modulated at a different frequency, yet all of these modulated frequencies are on one coaxial cable. A demodulation system is required at the end terminal (TV tuner). Broadband is more complex and more expensive than baseband transmission systems.

TELECOMMUNICATIONS TRANSMISSION METHODS

Not all point-to-point or point-to-multipoint telecommunications are the same. This section covers different types of transmission methods available today for voice, data, and video.

Asynchronous Transmission

The simplest synchronizing method is asynchronous mode which is based upon a stop-start method. Asynchronous signals are always at the 1 level while no information is being transmitted. A stream of characters is preceded by a start bit at the 0 level followed by 8 data bits and then a stop bit at the 1 level. The major drawback to asynchronous mode is the amount of overhead information required. Overhead information in asynchronous transmission is the stop and start bit data that carries no information but is required for successful transmission. An asynchronous mode of transmission is sometimes referred to as a character mode of transmission (Figure 3.15).

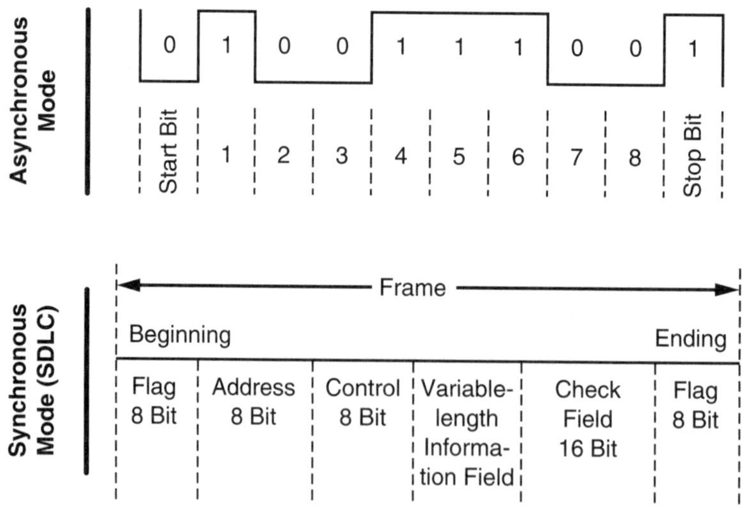

Figure 3.15
Asynchronous and synchronous transmission.

Synchronous Transmission

To keep transmission overhead at a minimum, data can be transferred in synchronous mode. Synchronous data is in a block format with information sandwiched between header and trailer records. A clock signal from the transmitter keeps both the transmit side and the receive side in synchronization. The major drawbacks of synchronous mode are the lack of standards and the complex transmission methods. Error correction techniques can be designed into the synchronous transmission format (Figure 3.15).

Note that many equipment manufacturers have their own synchronous proprietary standards, which are incompatible with recommended synchronous standards such as SDLC and HDLC.

Asynchronous mode terminals can always communicate with each other if the speeds of the transmitter and receiver are identical. Synchronous mode terminals require a master clock and compatibility or intelligence in the data terminal equipment (DTE). However, the advantages of greater data throughput and error correction techniques far outweigh the simplicity of asynchronous mode format.

Simplex Transmission

Simplex transmission is unidirectional transmission. A good example of this is broadcast radio or television. Transmissions are in one direction only. You cannot communicate back to the radio or television studio (Figure 3.16).

Half-Duplex Transmission

Half-duplex transmission is bidirectional. However, only one transmission can occur at a time. A good example of this is a walkie-talkie. Voice transmissions can only occur when the "talk" button is pressed. The distant end or receiving walkie-talkie station must be in "receive" mode while the originating walkie-talkie station is in "talk" or "transmit" mode. Once the originating walkie-talkie station has stopped transmitting, the distant-end station can then begin to transmit to the originating walkie-talkie. If both stations are transmitting simultaneously, neither party will receive the other's transmission (Figure 3.16).

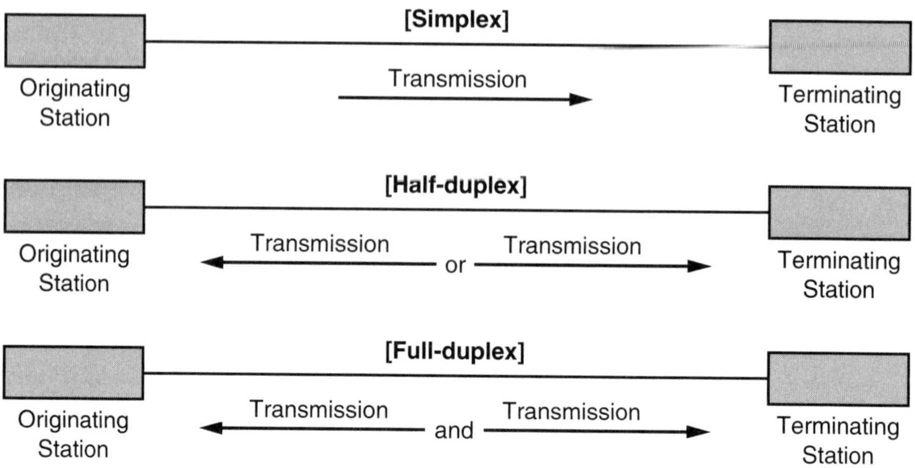

Figure 3.16
Simplex, half-duplex, and full-duplex transmission.

Full-Duplex Transmission

Full-duplex transmission is bidirectional. Both originating and terminating stations can transmit at the same time. A good example is a PSTN telephone call. Both originating and distant-end telephones can "talk" at the same time (Figure 3.16).

TELECOMMUNICATIONS SIGNAL COMPRESSION....

Compression standards have become common when digital telecommunications information is transmitted. Why? Well, the desire is to squeeze as much information out of a given bandwidth as possible. To do this, we must reduce or compress the original signal to its essential information before transmission. After the information is received, we can then increase or expand the information back to its original form.

Compression and bandwidth work hand in hand. Without compression, the bandwidth requirements for transmission may be too great. With compression and limited bandwidth, you can compress a signal only so much without signal degradation, so a certain amount of bandwidth will always be needed.

In the mid 1980s, it was determined that audio and video transmissions could be digitized. Since bandwidth requirements for both audio and video were high, it was determined that a set of compression standards would be needed. Each type of transmitted signal, whether voice, video, or data, would require compression to reduce the number of bits needed for transmission.

The hardware/software tool used to achieve compression is called an "encoder"; the playback hardware/software tool is called a "decoder."

Let's examine the four basic types of transmissions in some detail: voice, full audio (music), data, and video.

TIMETABLE OF HISTORIC DIGITAL COMPRESSION EVENTS

1988	Moving Picture Experts Group (MPEG) working group 11 is formed.
	MPEG-1 audio and video standard is created.
1990	MPEG-2 broadcast video standard is created.
1993	MPEG-3 standards are dropped in favor of MPEG-4.
1996	MPEG-4 working standards for intelligent audio and video compression are in progress.
1999	MPEG-4 standards for intelligent audio and video compression are adopted.

Voice (Speech) Compression

Recall that earlier in this chapter we reviewed the DS-0 level, which provides uncompressed voice at 64 kbps. The 64 kbps is derived from 8,000 samples/second × 8 bits/sample, or 64 kbps. Significant development effort has been spent on reducing the number of bits required to transmit voice. The results of this research have led to the development of compressed voice algorithms, which can provide a reasonable quality of service comparable to uncompressed voice. Compressed voice can operate at 4.8 kbps, 9.6 kbps, and 12.6 kbps, which is a significant reduction from uncompressed 64 kbps. A new compressed voice standard is under development; it will operate at 2.4 kbps. All of these reduced speeds involve an approximation of complex speech waveform frequencies.

Let's examine the components required to build a compressed voice encoder. First and foremost, the analog waveform must be converted to digital. The digitized speech signal is broken into its constituent building-block sounds, called phonemes. Each phoneme is recognized, and a variable-length code for the phoneme is looked up in a codebook list. The matched code is placed in buffer memory and then sent as a serial bitstream (Figure 3.17).

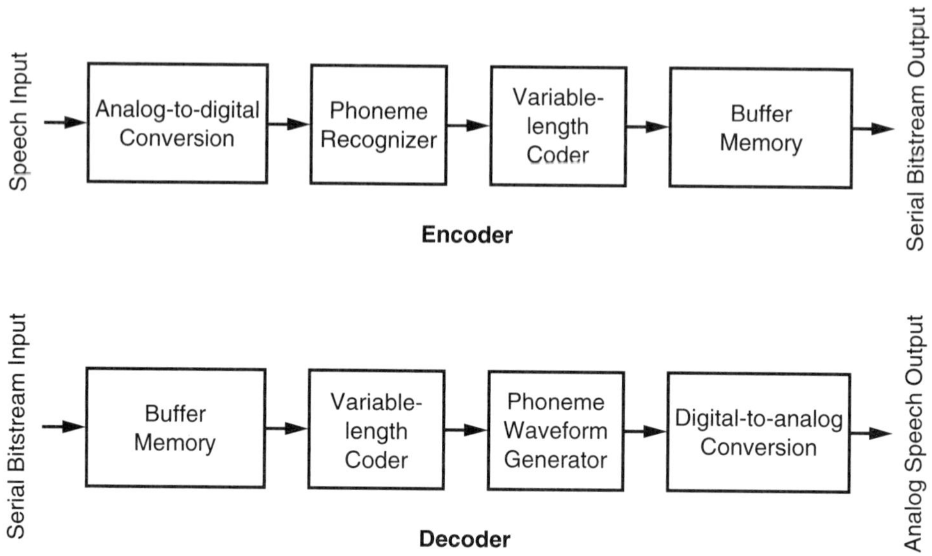

Figure 3.17
Voice compression encoder and decoder.

In the compressed voice decoder, the serial bit stream is fed into the variable-length decoder from buffer memory. The output of the variable length decoder selects a phoneme waveform from a predetermined list and sends the information to the digital-to-analog converter for conversion into an analog waveform.

There are approximately 100 phonemes in the phoneme alphabet. Utilization of phonemes, variable-length codes, and bit reductions can provide up to 1/1,700 of the uncompressed DS-0 bit rate.

It is evident that regardless of who is speaking, the output of this decoder is always the same. In other words, the phoneme waveform generator determines the output of the decoder. To compensate for this dilemma, personalization layers are added. MPEG-4 coding of audio provides techniques for representing natural sounds (speech and music). Code-exited linear-predictive (CELP) MPEG-4 coding tools provide functionalities specifically designed for speech. Therefore, compressed voice can operate at 4.8 kbps, 9.6 kbps, and 12.6 kbps with outstanding quality approaching uncompressed voice quality at the DS-0 level.

MP3 (Audio) Compression

MP3 is an audio compression technique. Uncompressed digital audio files or .WAV files are very large, consuming 10 megabytes of bandwidth or more per song. MP3 compressed files may consume approximately 1 megabyte of space per minute of audio, thereby reducing the size of the digitized audio file. Because of the relatively small size of MP3 compression, audiophiles can easily trade music over the Internet. MP3 compression has also enabled audiophiles to store huge music collections on their computer hard drives and play back selections in any order at any time.

MP3 does not produce a size reduction by traditional compression methods. The audio stream is analyzed and compared with models of human hearing and perception. The audiophile can configure the amount of data to be preserved by providing the optimal balance between file size and audio quality.

Any audio source contains a tremendous amount of audio information that is never even heard or that the brain can even process. MP3 compression seeks to discard information, such as very low or very high frequencies, that cannot be heard.

MP3 uses the MPEG-1 layer III encoding-decoding scheme, which relies on human psychoacoustics and traditional compression methods. For example, the brain can perceive two frequencies or notes as one frequency or note. Another example is a loud and soft frequency or note could be perceived as one loud frequency or note. This acoustical data can be represented in tables and charts and then converted to mathematical models representing human hearing patterns.

MP3 compressed files are composed of a series of relatively short frames with a header that contains extra information about the data coming in the next frame. These files also contain data about the song, title, artist, etc. This is called the ID3 data; it is usually stored at the beginning of an MP3 dataset. Future releases of ID3 may include security.

As we mentioned before, the quality of audio is determined by its sample rate, which can be controlled by the audiophile. Sample rate is a measure of how many times a second the audio stream is digitally represented (reread "Quantizing and Companding" in this chapter). We already know that uncompressed CD-quality audio is sampled at 44 kHz. MP3 compression can be set at 64 kbps, 96 kbps, 128 kbps, 192 kbps, etc., sampling rates. 128 kbps is considered near CD-quality sound. Of course, the higher the bit rate, the better the quality. Also note that MP3 quality cannot improve the original audio source information. The function of MP3 is for compression only.

Some MP3 recorders have a variable bit-rate (VBR) capability. VBR improves sound quality while maintaining a relatively low file size. This is done by varying the sampling rate while recording. For example, an MP3 file recorded with VBR at 128 kbps can sound as good as a 192 kbps MP3 file without VBR.

Creating MP3 files is generally a two-step process. You can extract audio from CD format into an uncompressed format or .WAV format and then run the uncompressed file through an MP3 encoder. However, software "rippers" may allow you to encode from CD to MP3 in one single pass.

MP3 compressed files may have a significant impact on the telecommunications networks of the future and the distribution of music through Internet streaming audio.

MPEG-1 and MPEG-2 (Video) Compression

Uncompressed analog video is an inexpensive, effective method of transmission. However, it is very wasteful of bandwidth. A 6 MHz uncompressed video signal contains considerable redundancy in the image. For example, a 6 o'clock newscast may have an anchorperson sitting at a desk reading the news from a teleprompter. When video compression is used, the transmission will consist of only the changing portions of the image, which is minimal, compared with a transmission of the entire image.

The uncompressed video signal therefore contains considerable redundancy. Often, large amounts of information consist of a background image, which does not change from frame to frame. Removing this redundant information can compress the video signal to a reasonably narrow bandwidth.

Considerable efforts have been made to send "narrowcast" video through DS-0 levels. The quality at DS-0 level falls short even for video conferencing. Generally, excellent full frame, full motion compressed video can be obtained at 512 kbps. Most video conferencing services settle on 384 kbps or 6 DS-0's as a reasonable balance between quality and cost. In many cases, bonded BRI-ISDN lines totaling 384 kbps can be used for videoconferencing.

A block of data can be transmitted only with the changed information. Otherwise, the decoding equipment retains the previous block and forwards it to the receiver. This practice is also known as *predictive coding.* Predictive coding analyzes the elements that are changing and predicts what the next frame will be. Approximately every two seconds the entire frame is transmitted. However, between these frames, only the changed information is transmitted (Figure 3.18).

The Moving Picture Experts Group (MPEG-1) standard was developed as an audio/video standard. However, the audio and the video can be used separately or together. MP3 standards were developed from the MPEG-1 audio model. MPEG-1 was the first attempt to provide digitally compressed, synchronized, audio/video for telecommunications transmission. MPEG-1 audio/video is optimized at 1.5 MHz band-

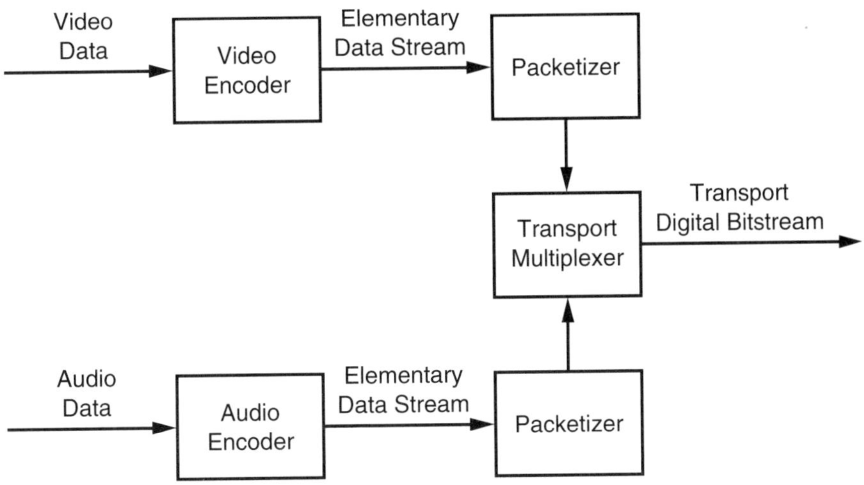

Figure 3.18
MPEG-2 encoder.

width; MPEG-2 audio/video will operate at virtually any bandwidth. MPEG-2 standards were developed specifically for broadcast transmission, and MPEG-1 was developed for compression standards.

While videoconferencing is costly and limited today, it is expected to become the standard method of telecommunications in the future.

We further examine uncompressed broadcast quality video signals in the next chapter. We also examine MPEG-2 digital satellite broadcast transmission in Chapter 4.

MPEG-4 (Video) Compression

Much of today's video streaming via the Internet uses intelligent MPEG-4 compression. MPEG-4 compression is a 10- to 100-times improvement over MPEG-2. Dramatic increases in coding efficiencies are available with MPEG-4. In addition to coding efficiency, many other services are provided by this new standard.

The basic building block of MPEG-4 is the scene—each scene comprises a set of objects that can be combined to recreate the scene. The objects can be general in nature; for example, a background scene, a person's head, etc. The units of objects are called routers and modems: video objects (VOs). A VO can be a talking person that can be joined with other objects, such as a background, to create a scene.

The MPEG-4 video coding algorithms provide an efficient representation of each of the VOs. The VOs by themselves use MPEG-1 and MPEG-2 compression techniques. The advantage of MPEG-4 is that the compression efficiency can be significantly improved for some video sequences by use of appropriate "tools" for each of the VOs.

While all of this discussion of coding techniques, VOs, scenes, and algorithms may be overly complex, it is important to note that MPEG-4 development is in its early stages and will continue to influence both audio and video compression techniques for the next several years.

Data Compression

Data compression has become a standard feature on many devices, including modems. Data is compressed to reduce or eliminate redundancies or repeated information found in the data. The data encoder algorithms detect strings of characters and represent them as a token. Similarly, the decoder works in reverse, enabling the restoration of the original data.

Compression ratios can differ for each application. The compression ratio may be 6:1 for file transfers, whereas sporadic query data may provide a compression ratio of less than 2:1. For example, a 56-kbps dial-up line can deliver 112-kbps to 256-kbps or higher throughput when data compression is applied. Throughput is the number of information bits that are transmitted successfully per unit of time. Overhead and signaling bits are not analyzed when throughput is examined.

Data compression is also important for transmission systems over long distances where cost may be a factor. Compression of the information prior to transmission can mean that less bandwidth will be needed, ultimately reducing cost. Data compression can also extend the life of a DS-0-level (64 kbps) private dedicated line, where it may be too costly to upgrade.

CONCLUSIONS

Digital transmission has revolutionized the telecommunications industry. Today's transmission systems are faster, cheaper, and smarter than their old analog counterparts. Errors occur much less frequently, and compression algorithms have improved. All of these benefits allow much more data to be sent in the same amount of bandwidth.

We also looked at different network configurations and dedicated transmission systems. We reexamine these in more detail in Chapter 7 when we take a closer look at requirements for the small business office.

4 Telecommunications and Television Services

The telephone has certainly had an impact on our lives and the lives of previous generations. Certainly, the telephone will continue to impact the lives of future generations to come. The concept of universal telephone service has provided us with the luxury to communicate from virtually any place at any time.

Television, however, has become a genre in itself. More U.S. households have television than have a telephone. It is one of the most successful technological achievements of our time. Both the video (picture) and audio (sound) are simulcast by guided (cable) or unguided (airwave) media. Presently, television is a one-way transmission providing news, weather, sports, and entertainment within our homes.

Television or TV technology may be changing within the next few years with the advent of interactive television and high definition television (HDTV). We discuss more about HDTV in Chapter 10.

So what exactly is standard definition analog television (SDTV)? This chapter covers cable, satellite, and standard television services over the airwaves.

The first section covers early development of television and the race to provide these services.

The second section covers television technologies, describing how both black and white and color television work and what information a video signal contains. We also take a brief look at projection television systems.

The third section looks at recorders and players that are integrated with television: specifically, the VCR, PVR, and the DVD.

The fourth section of the chapter contains specific information regarding analog cable television services.

The fifth and final section includes information regarding digital services for television, including satellite TV.

Hey Ma,….Look "whutz" on the TV!...

THE RACE FOR TELEVISION SERVICES..................

The invention of television was a group effort. No single inventor or company was credited with the discovery of television. Instead, modern-day television standards have evolved from different inventions in the early 20th century.

Early Television

Philo T. Farnsworth first experimented with the transmission of electronic television signals. In 1928, he was able to transmit crude television signals for over one mile to viewers at the San Francisco Merchant Exchange building.

Early television cameras, called iconoscopes were designed to pick up information and convert moving images into hundreds of separate horizontal lines. These early systems, developed by Vladimir Zworykin, were experimental, with low resolution. Zworykin initially worked for Westinghouse on radio projects but could not resist the lure of television. In 1929, Zworykin invented the all-electronic camera tube, called the kinescope, which serves as the basis for all television tubes in use today. David Sarnoff, then the president of Radio Corporation of America (RCA), eventually hired Zworykin to develop his television system for RCA.

Other companies such as EMI and Marconi were in a race to develop television patents first. RCA eventually won out and presented its television system at the 1939 New York World's Fair.

In 1939, The National Broadcasting Company (NBC) started broadcasting to 400 sets in the New York metropolitan area, using 340 scanning lines at 30 frames per second. Three years later, in 1941, the Federal Communications Commission (FCC) established standards for television broadcast services. The FCC and the National Television Standards Committee (NTSC) set these standards at 525 lines of resolution at 30 frames per second. These standards are still in use today.

After the war there was a two-year delay in the development and manufacture of television receivers while color television services were explored by the FCC. It was decided that black-and-white sets would be manufactured since the development of color television technology was not advanced enough at the time.

TIMETABLE OF HISTORIC TELEVISION EVENTS

1873	British scientists W. Mith and J. May discover photoconductivity used in camera tubes.
1878	Sir William Crookes confirms the existence of cathode rays.
1897	Karl Braun of Germany invents the cathode ray tube (CRT).
1907	Lee De Forest invents the triode electron tube. This made amplification of video signals possible.
1923	Westinghouse, General Electric, AT&T, and RCA begin television research.
1923	RCA inventor Vladimir Zworykin patents the electron camera tube iconoscope.
1927	First long distance television via AT&T facilities is broadcast from Washington to New York.
1928	First television via wireless transmission facilities is broadcast in San Francisco.
1929	Vladimir Zworykin demonstrates electronic television camera/receiver.
1930	NBC is granted an experimental broadcast license.
1939	RCA displays television at the New York World's fair.
1941	NBC and CBS obtain commercial broadcast licenses.
1945	An estimated 10,000 television sets exist in the United States.
1952	An estimated 22 million television sets exist in the United States.
1953	An NTSC television standard is adopted in the United States, allowing for color transmission compatible with the monochrome standard.
1954	Commercial color broadcasting using the NTSC standards begins in the United States.
1955	Videotape is introduced for storage of video and audio.
1960	An estimated 60 million television sets exist in the United States.
1962	Telstar 1 satellite is launched for television satellite broadcast transmission.
1965	RCA displays color television at the New York World's fair.
1965	NBC begins broadcasting all programming in color.
1974	97% of American homes have one TV set running at least 5 hours a day.
1984	Stereo television broadcasting begins.

Television Standards

In 1954, The NTSC set the standards for color broadcast television services. The standard chosen was compatible with existing black-and-white standards. This was important since existing television sets would not be obsolete. The color picture would be obtained by inserting color luminance signals inside the black-and-white signal.

Other color television broadcast standards are in use outside the United States. SECAM (sequence with memory) and PAL (phase alteration by line) are used extensively in Europe and are not compatible with the NTSC standard. Each of these systems represents 625 scanned lines per video frame with a rate of 25 frames per second. All of Europe, the Middle East, Asia, and Africa are using the SECAM or PAL video standard.

In the 1970s, television became the primary source of news and information in the United States. Today, it is estimated that over 700 million television sets are used worldwide.

TELEVISION TECHNOLOGY

The analog television set has been the cornerstone of electronic components in the household for several generations. The basic TV technology hasn't changed in over 50 years! So why the sudden rush to eliminate analog TV and move to digital TV and HDTV? We must understand analog television signals first before we explore digital television in some depth.

Since the advent of personal computers in the 1980s, television clarity and resolution aren't nearly as good as that of computer monitors. An analog TV can display up to 525 lines of resolution every 1/30th of a second. The lowest computer monitor, however, displays 640 x 480 pixels. A pixel is considered a dot on the screen. When comparing a low-end computer monitor with a television set, we can say that the TV displays 512 x 400 pixels vs. the monitor at 640 x 480 pixels. There is simply no comparison.

The television signal comprises an amplitude-modulated (AM) video signal and a frequency-modulated (FM) audio signal. In an unguided system, the television signal is broadcast from the highest building or tower in an urban area. The reception of the analog television signal will be improved the closer the television set is to the transmission tower or building.

Let's examine the analog TV receiver in greater detail.

Television and Human Interaction

Remember the days of the flip book? You drew a figure on a piece of paper and then varied the figure from one page to the next. You could make the figure change direction, bend up and down, walk or move from one side to the other by flipping the pages. Well, television works on these same principles. If you divide a movement into a series of still images, you can conversely reassemble these still images into movement by changing the images in rapid succession. This technique relies on a phenomenon associated with the human brain.

Another interesting principle about television and humans is the concept of a pixel. Pixels are the individual dots on a television screen which, when connected, make up a picture. Television, computer screens, newspapers, magazines, etc., rely on pixels or small colored dots for an image. The human brain again fuses these individual pixels and creates an image.

A third phenomenon is that the brighter the still image presented to the viewer, the shorter the persistence of vision. Therefore, if the space is longer than the persistence of vision, the image flickers. For reduction or elimination of "flickering," two picture "flashes" are created. In this technique, known as interlacing, the picture or image is scanned twice. Interlacing is covered in the Television Picture Scanning section of this chapter.

If you stand very close to a television screen, you can begin to see the individual pixels making up the image. If you stand at the recommended distance from the television screen, however, you perceive the pixels as merged and your brain no longer interprets them individually.

Television and the Cathode Ray Tube

The most important part of a television receiver is the cathode ray tube (CRT). The CRT converts electrical signals into electron beams of varying intensity. The electron beams are scanned from left to right across the picture tube creating a scan line. All types of television use this concept of electron scanning.

At the transmission end or television studio, a television camera tube (iconoscope) was once used to scan and pick up the light or dark areas of a picture and convert them into electrons of varying intensity. Today, electronic solid-state devices can be used instead of a picture tube to accomplish the same goal. In spite of recent advances, the CRT is still the most popular display device even with solid state and plasma devices gaining in popularity.

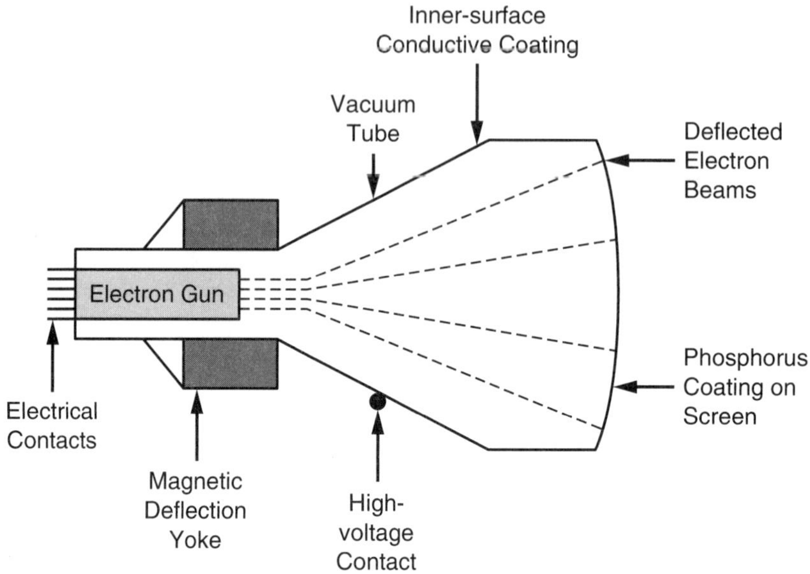

Figure 4.1
Cathode ray tube.

The CRT contains an electron gun (at the back of the tube). This electron gun constantly emits electrons past a control grid that regulates the strength of an electron beam. A high voltage contact within the tube causes the electrons to move toward the glass faceplate screen. The electron beam is then deflected by a magnetic field, or yoke, and passes a control grid that causes the beam to vary in intensity (black or white). The beam is then focused onto a small spot on the glass faceplate, which is capable of glowing (Figure 4.1).

A phosphor is used on the glass faceplate. This material, when exposed to an electron beam, emits a glow. In a black-and-white screen, there is one white phosphor that glows when struck. There are many upon many different phosphors, characterized by the length of time an emission will last.

Television Picture Scanning

The electron beam literally paints an image onto a screen one electron beam at a time. The beam is moved from one side of the screen to the other by magnetic coils that move the beam in a raster scan from one line to the next. When the beam scans from left to right, the intensity of the beam is modified to create different shades of black

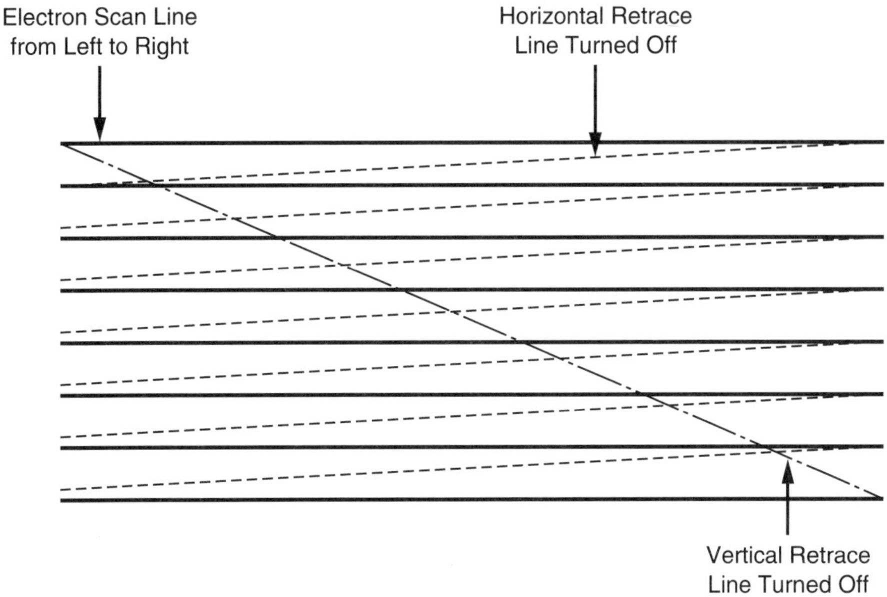

Figure 4.2
Television scanning.

and white. In Figure 4.2, the solid horizontal lines represent pixels that the beam is painting. The dashed lines represent the beam flying back from right to left to perform the next scan. It is important to note that the electron beam is turned off during the time the beam needs to move from right to left to begin the next scan. This movement is referred to as the horizontal retrace (Figure 4.2).

Vertical retrace signals move the beam from bottom right to top left.

All television screens use a technique called video interlacing when painting the screen. Only half of the lines are painted per frame. The beam paints only every other line (odd). After completing a screen, the beam then returns to paint the remaining line (even), which completes the picture.

Computer monitors with high resolution generally do not use the video interlacing technique because of flicker. Instead, progressive scanning is used; it paints every line on the screen in sequential order.

For television, the electron beam literally paints 525 lines 30 times per second or a total of 15,750 lines per second.

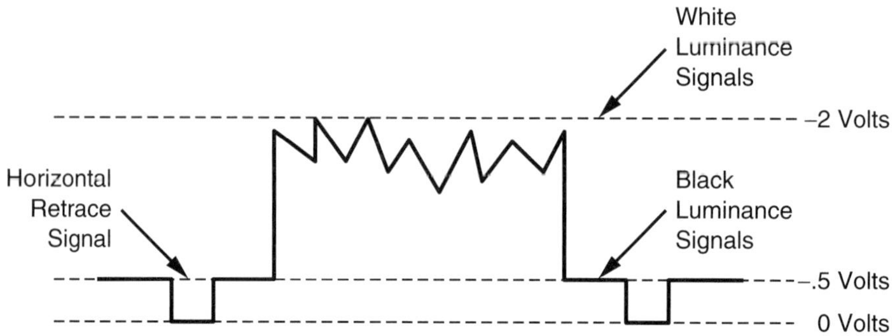

Figure 4.3
Television signals.

The television receiver needs to synchronize with the television broadcast station so that the electron beam will precisely draw the correct picture. Horizontal and vertical synchronous pulses perform this function.

A television signal that contains the vertical retrace information, horizontal retrace information, and the intensity information without the broadcast transmission carrier wave and the audio information is called a composite video signal. It is usually an RCA jack (colored yellow). The composite video waveform can be viewed on an oscilloscope (Figure 4.3).

Analog (Monochrome) Television

Standard definition (monochrome) television (SDTV-B&W) began broadcasting via the airwaves in 1941. In the United States, the FCC agreed upon a standard typically referred to as an NTSC signal (NTSC is the national television systems committee). SDTV is transmitted in analog format with the video amplitude modulated while the audio is frequency modulated.

SDTV-B&W is essentially the same black-and-white television set used in 1941. The technology has successfully evolved from vacuum tubes to transistors and from transistors to integrated circuits. With these advances in the technology, the television set has migrated from a heavy stand-alone device to a miniaturized device not much larger than a thumbnail.

The SDTV analog format relates the strength of the transmitted signal to an inversely related video. In other words, the darkest portions of the video signal have the most transmitted power, and the lightest portions of the video signal have the least transmitted power.

The bandwidth of an analog television signal is 6 MHz. The sound carrier is 4.5 MHz off the carrier frequency for the monochrome information. There is also a gap of 1.25 MHz on the low end and .25 MHz on the high end to avoid crosstalk with other channels. Note that basic broadcast television at very high frequency (VHF) consists of 12 channels (Table 4.1).

Table 4.1 Television Frequency Allocation

Frequency (MHz)	Channel	Frequency (MHz)	Channel
54–60	2	180–186	8
60–66	3	186–192	9
66–72	4	192–198	10
76–82	5	198–204	11
82–88	6	204–210	12
174–180	7	210–216	13

Analog (Color) Television

In the late 1940s, the Columbia Broadcasting System (CBS Inc.) initially developed a 405-line color TV system that used a frame rate of 24 frames per second. This system was incompatible with the NTSC black-and-white systems developed earlier. The RCA color system, developed after the CBS system, was compatible with the NTSC black-and-white standard. This made color television backward compatible.

The RCA solution was to use a secondary subcarrier frequency, embedded within the monochrome TV, or "luminance," signal to carry the color, or "chrominance," information.

The selection of colors for the display was very important because of the small range of available phosphors. The three colors associated with television are red, green, and blue. With all red, green, and blue turned on, you obtain white. With all red, green, and blue colors off you obtain black. When you mix these colors in different variations, you obtain a mixed color.

The transmitted signal for color television requires a color synchronization signal called the "color" burst. A subcarrier frequency of 3.58 MHz is inserted into the horizontal blanking signal. A subcarrier frequency is a secondary carrier frequency superimposed upon the carrier frequency (Figure 4.4).

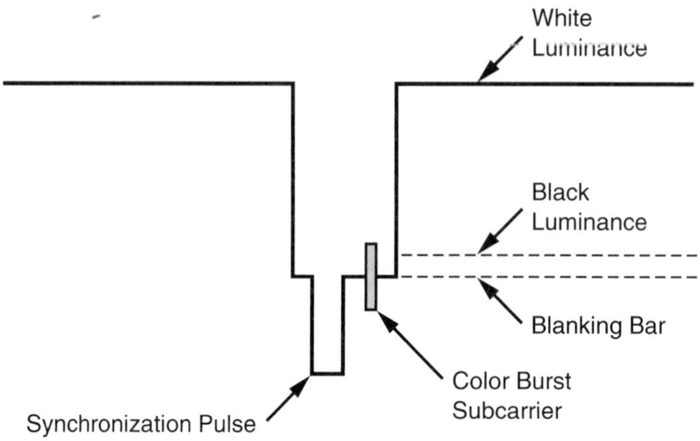

Figure 4.4
Color television transmission signals.

The color television CRT is somewhat different from a conventional black-and-white CRT. Instead of one electron scan, three electron scans move simultaneously across the screen. These three electron scans are associated with the primary colors of television: one beam for red, another for blue, and another for green.

The screen itself contains red, blue, and green phosphors, as opposed to black-and-white screens, which contain a single phosphor for the entire screen. In a color receiver, the red, blue, and green phosphors are arranged in stripes (or dots). Each of the three electron beams is designed to vary in intensity in accordance with the amount of red, blue, or green necessary for each pixel on the screen.

Between the screen and the electron gun is a thin metal screen called a shadow mask. The mask aligns the electron beams precisely on the red, blue, or green stripe without any shadow or unnecessary mixing of colors (Figure 4.5).

Projection Television

In recent years television screens have become larger. The demand for larger screens requires a projected image. CRTs can be manufactured up to 40 inches diagonally. Therefore, any screens greater than 40 inches must be projected onto a screen. Television projection receiver devices can be either rear projection or front projection systems.

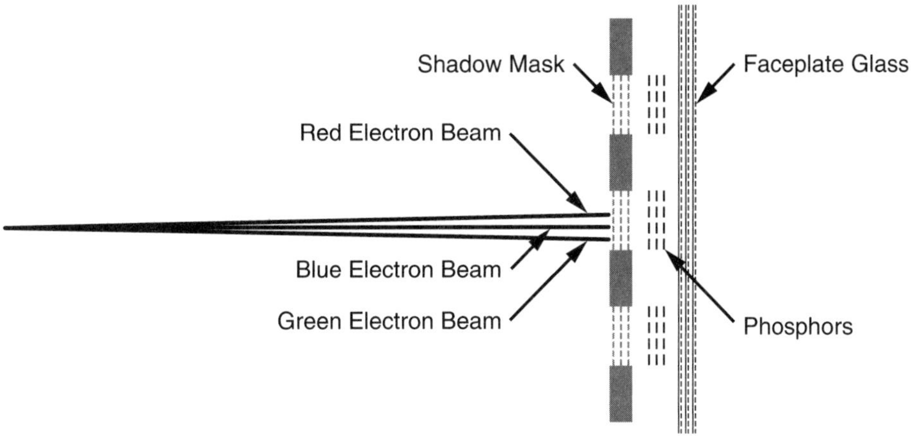

Figure 4.5
Color television CRT.

Rear projection television receivers are self-contained. The plastic screen is built into the unit, and the projection lens is behind the screen. The television viewer sits in front of the plastic screen to view the image. Front projection television systems are designed like a movie theater where the projected image is focused on a reflective screen that is in front of the viewer.

A CRT is generally not used in a projection system. Instead, a solid state microdisplay with pixels is used instead of a conventional electron scan. The individual dots or pixels have a variable intensity and use the primary colors of television: red, blue, and green.

Two types of projection system designs are available today. The most basic devices contain a microdisplay inside the projector. An intense light shines through the microdisplay and projects the image onto a screen. This is called transmissive projection. Another approach is to shine an intense light through the microdisplay, reflect it off a mirror, and project the image onto a screen. This is called reflective projection (Figure 4.6).

Advances in projection television technology have led to sophisticated devices that are small, lightweight, and have higher resolution than standard NTSC television can provide. Therefore, it is only natural that video projection systems have migrated to the computer for presentations, lectures, and conferences. Most projection devices can handle either composite video or video from a computer.

I'll provide you with more about projection video and the computer later in the book.

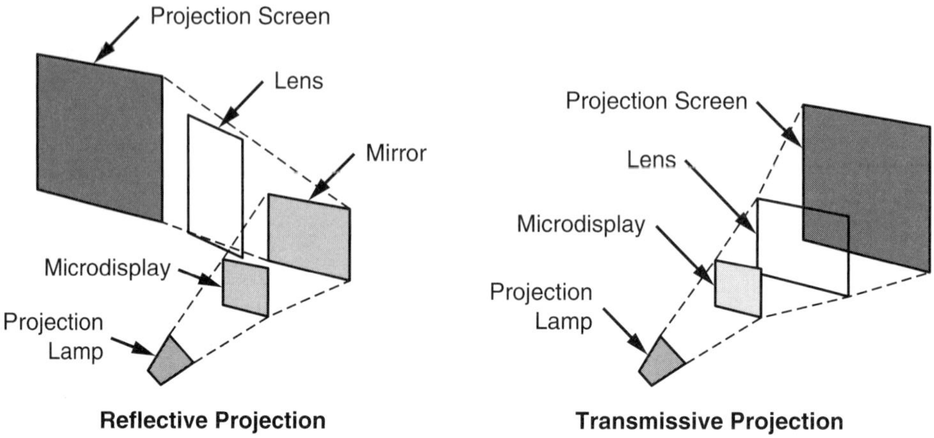

Figure 4.6
Projection television.

VIDEO RECORDERS AND PLAYERS

You cannot complete a discussion of television without analyzing video recorders and players. Unlike television technology, which has remained relatively stable for the last 50 years, video recording devices have undergone a migration path from analog to digital.

The first recording device, the VCR, is as popular as a television receiver. The VCR, an analog device, stores video information such as movies and television programs on videocassette tape. Many television receivers are a combination of TV and VCR, providing consumers with the ability to both record and watch television programming at the same time.

Newer devices such as the personal video recorder (PVR) and digital versatile disk (DVD) use digital formats to store and retrieve video information in a format similar to that of the VCR.

VCR Technology

The VCR, created in the 1970s, allowed individuals to gain control of television programming by providing a method of recording for later playback. The VCR created the video rental store and a large marketplace for movies and television shows to be viewed over and over again.

Timetable of Historic VTR and VCR Events

1951	Ampex examines a rotating-head VTR approach.
1954	Amplitude-modulated (AM) signals for a VTR are used on an Ampex experimental machine.
1955	Ray Dolby simplifies Ampex VTR electronics by using frequency-modulated (FM) signals.
1956	The Ampex VTR, called the Mark IV, is first demonstrated.
1956	Videotape is first used on-air to delay "Douglas Edwards and the News" by three hours.
1961	A helical scan format VTR called the VR8000 is developed, improving quality once again.
1966	Sony experiments with the idea of tape guides pulling tape out of a cassette. JVC buys the patent and creates a VHS cassette machine.
1972	JVC creates an easy-editor for the model 900 VTR.
1973	The JVC consumer-based VHS VCR machine is introduced.
1973	The Sony consumer-based Betamax VCR machine is introduced.
1985	Blockbuster opens its first tape rental store.

VCRs are an intricately mechanical electronic device. There are a motorized tape loading and ejection system, motorized tape paths, a motorized rotating drum.

Half-inch-wide video tape (oxide-coated Mylar) is enclosed in a plastic shell with a door. The door is necessary so that the tape will not be touched by human hands. Several low-friction rollers and spring-loaded locks within the plastic shell allow the tape to move freely inside without unrolling.

The audio information is stored in a linear format on the tape. The video information, however, is stored in a slanted or helical format (Figure 4.7).

The slanted video information is recorded by two record/playback heads that are mounted on a tilted rotating drum. The heads are 180° apart. We already know that a television image is 525 horizontal scan lines that are interlaced. A pass of the first rotating head writes the image data for the odd scanning lines. When the second head passes the tape, the image data is written for the even scanning lines, completing a picture. Therefore, the drum must revolve once for both heads to write to the table, providing a full 525 scanning lines (Figure 4.8).

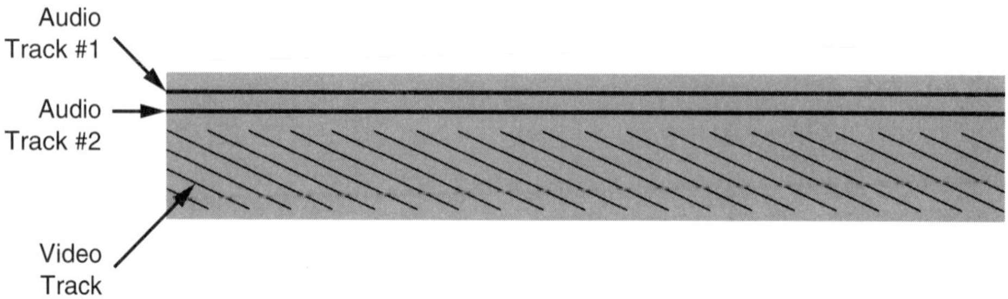

Figure 4.7
Video tape recording.

PVR Technology

Within the next several years, sales of the PVR are expected to outpace the VCR by a wide margin. The PVR device is popular because it provides all of the functions of a VCR, such as record, playback, fast-forward, rewind, and pause, with digital quality and reliability.

PVRs do not use videotape as a conventional VCR does. Instead, PVRs use a hard disk similar to a PC, along with compression techniques we discussed in the last

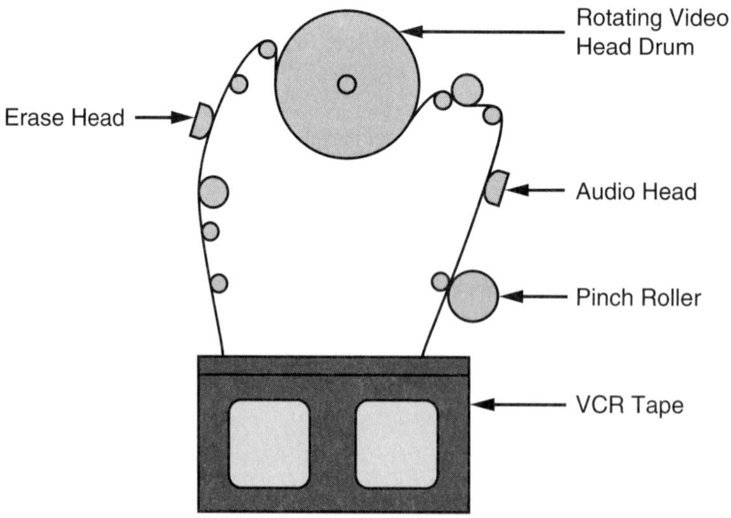

Figure 4.8
VCR mechanism.

chapter. The PCR is basically a stripped-down PC with a large hard drive for storage of video information.

PVRs have an MPEG2 encoder/recorder built within the device to compress the video signal to a smaller size before storing the digitized information on a hard disk. For playback, PVRs contain an MPEG2 decoder to expand the signal back to its original analog form for playback on a standard television receiver. Because it is a digital device, the PVR provides superior audio and video quality and virtually noise-free recording.

The PVR's hard drive is usually 10 gigabytes, which holds approximately 20 hours of MPEG2 encoded data. When the hard drive runs out of space, the oldest video program is removed automatically to provide hard-drive space for new programs.

Additional features of PVRs allow for the downloading of programming schedules through a modem, recommend television programming based upon consumer taste, and automatically record programs based upon earlier choices.

The output of a PVR contains analog composite video that can be directly connected to a standard television receiver's video jack.

The PVR behaves like an intelligent VCR!

DVD Technology

Digital versatile disk (DVD) players have been available since 1997. The DVD player decodes digitized MPEG2 format 5-¼-inch disks (double sided/double layer) containing up to 15.9 gigabytes of data. A DVD can contain over 8 hours of high-resolution video (double sided/double layer) with 720 scan lines of information.

Because there is so much room on a DVD, it has become common practice to add additional features such as commentary tracks in which the filmmaker or director explains how the film was made. DVDs can also contain extra unreleased scenes, subtitling, information services, and Dolby digital AC3 surround-sound audio.

You can purchase a stand-alone DVD player or your PC may include a DVD player.

Even though the DVD storage capacity is huge, the uncompressed data of a full-length movie would not fit on a DVD. This is why the data is compressed and encoded in MPEG2 format.

A laser assembly shines a laser beam onto the surface of the DVD that can read a pattern of bumps. The bumps can be either high or low, representing the 1's and 0's needed to process digital information. The laser precisely reads the information from

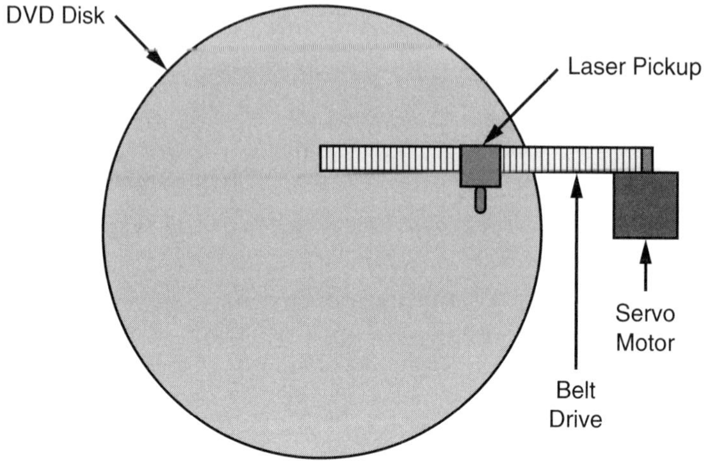

Figure 4.9
DVD player technology.

track to track as the disk spins between 200 and 500 revolutions per minute (Figure 4.9).

It is important to note that the rotational speed of the disk varies because of the laser's location when reading the disk. As the laser moves outward, the motor slows the spinning of the disk so that the information can be read at a constant rate.

If a DVD has a second layer, the start of that layer's data can be at the outside of the disk instead of the inside. In this fashion, the player can transition quickly from one layer to the next without delay or loss of quality.

The output of a DVD player contains analog composite video, which can be directly connected to the video jack of a standard television receiver.

The DVD player is similar to a CD player, which we discuss in Chapter 7.

CABLE TELEVISION

Cable television, or CATV, services are used in over 65% of U.S. households, with an available penetration of over 97%. The advantages of cable over standard airwave transmission are many. We discussed the advantages of guided media over unguided media in Chapter 1. Less interference and the provision of a cleaner signal are the most significant advantages.

TIMETABLE OF HISTORIC CABLE TELEVISION EVENTS

1948	Ed Parson of Astoria, Oregon, builds the first CATV using twin lead transmission from house-to-house.
1950	Bob Tarlton of Lansford, Pennsylvannia, uses coaxial cable on telephone poles for transmission from house to house.
1950	Martin Malarkey of Pottsville, Pennsylvannia, uses amplifiers and coaxial cable for cable transmission over longer distances.
1972	Home Box Office (HBO) begins to offer "pay per view" sporting events over cable transmission lines.
1975	HBO transmits the Ali-Frasier fight from a Manila satellite feed to Mississippi and Florida by cable transmission lines.
1980	Turner Broadcasting in Atlanta, Georgia, develops the first super-station concept of mini-networks distributed nationally by satellite.
1986	HBO begins encryption services, creating "pay" TV.

Cable television technology is made possible by the high bandwidth characteristics of coaxial cable. A television set owner utilizing an antenna can easily become a cable television set owner without any major investments in hardware or equipment. Moving the television input terminal from the antenna connection to the cable connection is all that is necessary.

Cable's success and high growth have attracted the attention of government regulators and legislators. Monopolistic business tendencies and public policy issues go hand-in-hand with cable television policies.

Cable television offers more programming than ever before. In addition to international broadcasts from satellite transmissions, cable also carries local broadcasts from high schools, libraries, or government offices. Users of cable television can be well informed on a global, regional, and local basis. Cable television also offers a variety of entertainment on diverse subjects such as cooking, rock music, sports, and history, and it supports children's programming and educational services.

Remember, cable television is a one-way transmission service. However, with recent demand for interactive television services and cable modem services, cable is slowly becoming a two-way medium.

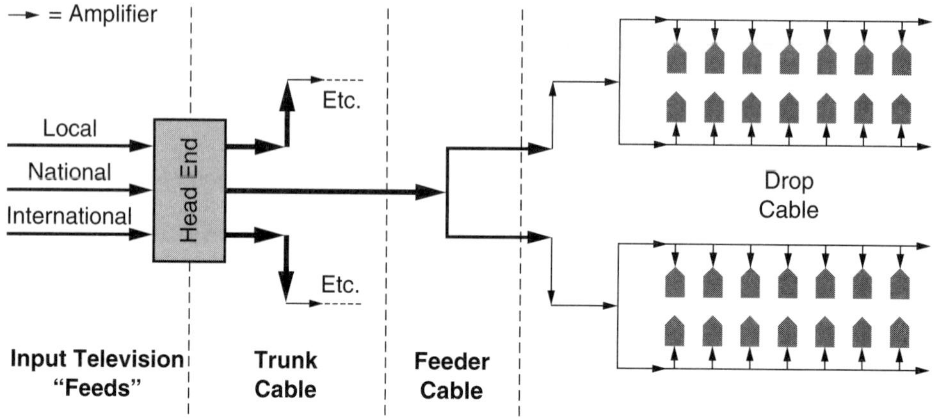

Figure 4.10
Cable television networking.

Cable Network Design

There are five major sections in a cable television system. The head end is the CATV distribution center. The head end is the center of all CATV activity where all of the programming is centralized and distributed. The second section of the cable television system is the trunk cable. The trunk cable is the major distribution artery to distribute programming and services. Today, cable trunks mainly consist of fiber. The third section of a cable television system is the feeder cable. The feeder cable is the neighborhood cable used to distribute programming locally. The drop cable is the fourth section of a cable television system. This cable provides access to the residence. The fifth and final section is the terminal equipment (television tuner, set-top box, etc.) for access to the specific television channel (Figure 4.10).

Early CATV systems used microwave transmission facilities to capture the signals from distant stations and then to "rebroadcast" the signals to a new location. This technique provided television signals in areas that were outside the normal viewing area. Cable TV became popular in rural areas initially because viewers could then receive programming that wasn't available before a cable system build-out.

In the early 1970s, cable TV became popular for different reasons. For the first time, additional programming channels were added to the normal allotment of television signals. Because cable TV used a closed transmission medium, the entire frequency bandwidth of the cable could be used for television signals. Cable designers used the frequencies between 88 MHz and 174 MHz for 13 channels of standard "airwave" programming and started new programming with channel 14 at 216 MHz. If the television device is "cable ready," a switch in the back of the set will provide all the additional channels that cable TV has to offer.

Even with a closed transmission system, interference can be a problem. "Airwave" broadcast signals can enter through connectors or from poorly designed coaxial transmission lines. Signal interference or "ghosting" can still occur on a cable transmission system, where the "ghost" can be a delayed airwave signal.

As the demand for more specialized programming increased, more channels were necessary. The earliest television systems generally had a bandwidth of 200 MHz, providing 33 channels. The available bandwidth was eventually increased to 550 MHz, providing 91 channels of programming.

Cable Television and Scrambling

Several analog scrambling systems have been used to prevent the reception of "unpaid" television signals over the last several years. One of the techniques is to remove the synchronization pulse from the transmitted television signal and then reinsert the signal at the customer's home if the service is paid for. Another technique is to insert a frequency that will interfere with the transmitted television signal and then remove the interference signal if the service is paid for.

In either case, the scrambled, televised video is unviewable.

Cable Television and Fiber Optics

In the 1980s, cable TV began using fiber optic technology in the trunk transmission lines. We already know that fiber optic transmission lines provide less interference with greater signal-carrying capacity. Fewer amplifiers are needed with fiber lines, thereby providing improvements in picture quality and system reliability.

In addition to quality and reliability, fiber optics also added another benefit. Fiber could provide a trunk signal to several hundred households. A single fiber cable could, therefore, provide local neighborhood programming and enhanced services such as the cable modem (covered in Chapter 5).

DIGITAL TELEVISION

In the early 1990s, cable television began experimenting with digital MPEG services. You will recall that MPEG compression squeezes video information into a much smaller bandwidth. Therefore, one digital MPEG transmission will occupy onc-tenth of one standard 6 MHz analog television channel. The benefits of digital transmission such as error correction and minimal interference are also present with the MPEG transmission format. Digital transmission through coaxial cable will be the wave of the future.

Also note that if we examine the entire bandwidth of a coaxial cable and use an MPEG format to transmit all channels, we can transmit up to 1000 channels of television on a single coaxial cable.

It is also interesting to note that analog scrambling methods do not apply for digital services. In a digital system, the television signal is encrypted. An encrypted signal must be unencrypted with the proper key, or a blank screen will result and the digital information will not be processed.

With the advent of audio and video compression in the early 1990s, digital television transmission is now a reality. Digital transmission with compression techniques allows for a reduced bandwidth. Digital television transmission can be received via cable (with a set-top box) or via satellite (with a decoder/receiver).

HDTV, which utilizes digital television transmission, is covered in Chapter 10.

Digital Satellite Television Transmission

In 1993, the DBS-1 satellite was launched from the French Guiana. The DBS-2 satellite was launched in 1994, and the DBS-3 was launched in 1995. Hughes Corporation launched these satellites for a direct broadcast television network. The combination of these powerful satellites, MPEG1/MPEG2 compression and low-cost sophisticated electronic decoder/receivers make direct television broadcasting viable.

These services were first launched in 1994 and now can boast a subscriber base of over 10 million subscribers. One of these satellite providers has a dish that offers 500 television channels from a single 18-inch dish.

Direct broadcast television uses one head end and an earth station dish as an uplink facility to transmit television signals to the satellite. These facilities are very sophisticated with a complexity exceeding several television stations and providing multiple channels. Each of these television channels must be digitized, compressed, and then encoded. Digital encoding requires a decoder at the receiver end to recreate the original signal.

You will recall from Chapter 1 that the satellite is merely a relay station in the sky. A direct broadcast television satellite (DBS) performs a similar function.

Also recall our discussion of digital services. Even though the satellite is thousands of miles away from the receiving dish, the picture will be clear and noise free, thanks to digital transmission technology.

Digital Television Satellite Home Equipment

The key to DBS television and its success is the relatively low cost of the consumer equipment. With service provider rebates, the cost of this equipment is sometimes $0. The price vs. quality issue is certainly designed to compete head-to-head with cable television.

At the receiving end, an 18-inch satellite dish receives the signal. This is a very small dish as compared with analog dishes, which can be 8 to 10 feet in diameter. As with any satellite signal, the dish must be aimed exactly at the satellite. Each satellite has a latitude and longitude position. The satellite's position and its relation to the earth must agree with both the horizontal and vertical pitch of the satellite dish. Consult with the dish manufacturer or programming supplier for exact dish alignment (Figure 4.11).

The satellite dish requires a low-noise block (LNB) that is focused at the dish. The LNB is slightly offset from the center of the dish to minimize signal blockage and is setup to receive signals as they reflect off the dish. This is why a satellite dish is sometimes referred to as a satellite reflecting antenna.

LNBs can be single or double. A double LNB requires two coaxial cables; a single LNB requires one cable. A double LNB is designed to connect to two receivers. In a home where more receiver/decoders are required, amplified satellite splitters can be used to split the signals even further and to add more receiver/decoders. It is important to note that each television set requires a separate receiver to view separate channels. The satellite television provider may require a separate fee to provide service for each receiver (Figure 4.12).

Because of the high frequencies used between the satellite dish and the receiver/decoder, RG-6 coaxial cable is used instead of standard RG-59 coaxial cable (used for standard cable television).

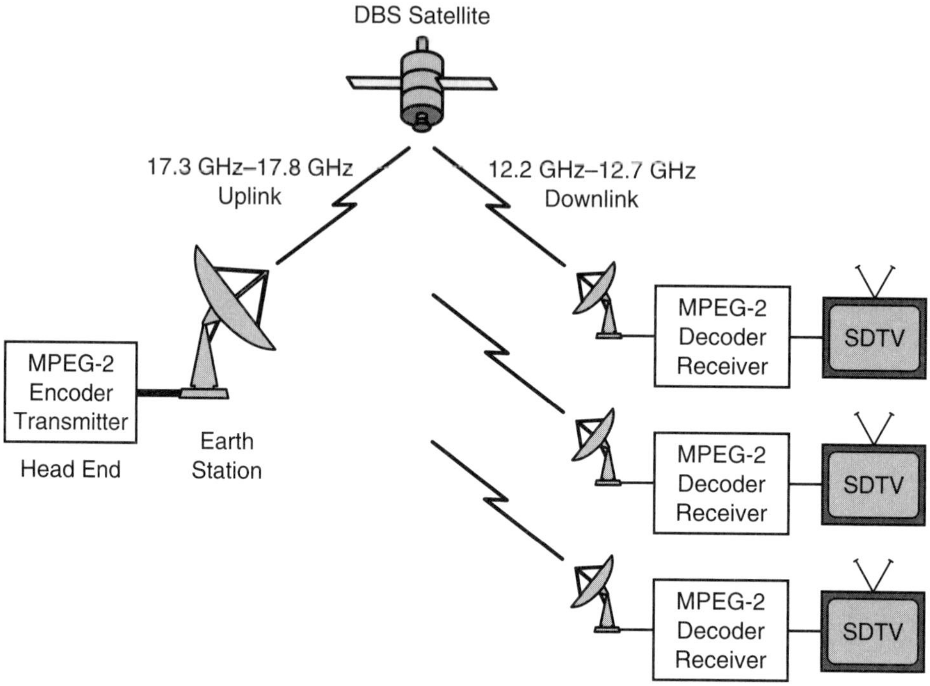

Figure 4.11
Direct broadcast satellite television.

Digital Television Encryption/Decryption

Both digital CATV and satellite TV incorporate encryption/decryption systems consisting of an encoder, an authorization center, and individual set decoders. The encoder converts the analog television signal to an MPEG-2 encrypted signal. This encrypted signal must be decoded before it can be displayed on a standard television set. The authorization center handles all subscription orders and controls all decoders within the system.

The Videocipher-rs (renewable security) encryption system can accommodate up to 256 levels of programming, with each level used to control access to a group of programs or a single program. This system has the capability of addressing up to 50 million receiver/decoders commonly called set-top boxes.

The Videocipher-rs system encodes video as well as the audio.

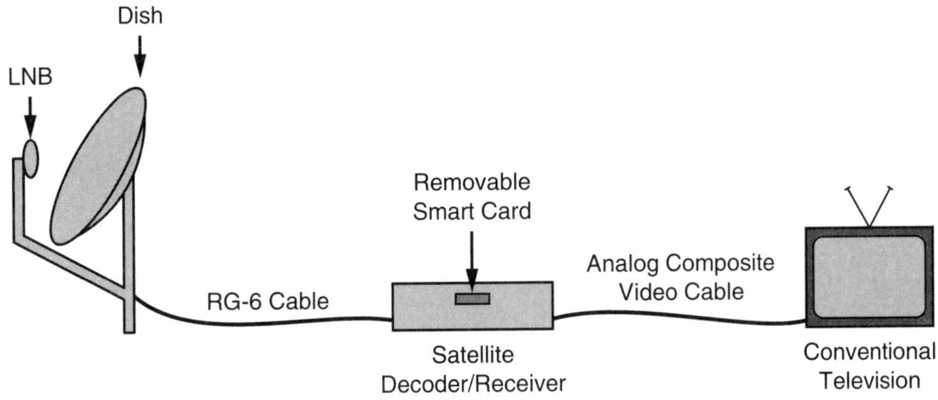

Figure 4.12
DBS satellite dish and decoder/receiver.

This system also features an electronic "smart card" that contains an integrated circuit. This circuit or smart card chip holds the mathematical algorithms that can decode television signals for each channel. It is very similar in functionality to an EPROM chip. Both smart cards and EPROM integrated circuits can be reprogrammed. This feature allows manufacturers and programming suppliers to update the set-top box's software or operating system.

Digital Television and Multiplexing

Compressed video can vary in transmission rate. MPEG2 can sometimes require as little as 1 Mbps and as high as 10 Mbps to create full-motion video. Unfortunately, if the transmission rate is set too low, picture quality will suffer. If the transmission rate is too high, bandwidth is wasted.

To resolve this problem, DBS systems generally use a process called statistical multiplexing. Recall our studies of multiplexing in Chapter 1. The analogy of a multiplexed circuit is to take the total bandwidth (apple pie) and cut the bandwidth into individual sections or channels (slices of apple pie). A statistical multiplexer constantly changes the individual sections or channels as necessary. This type of multiplexing allocates bandwidth according to need. Therefore, individual channel compression becomes a variable-rate bandwidth transmission system without sacrificing quality.

In this scenario, it is important to note that a situation can exist where all transmission channels require a higher bandwidth. If the transmission system, such as satellite, cannot accommodate the increased bandwidth requirements, all of the channels will suffer some loss of information and quality will temporarily suffer. Luckily, with a good transmission design, this will happen infrequently and when it does happen, the viewer will hardly notice the temporary loss of quality.

Set-Top Boxes

The rapid development of digital television services has led to the device called the set-top box. This device is no more than an MPEG2 decoder, a decryption section for security, and a receiver. The output of the set-top box is conventional analog composite video.

The architecture of a set-top box can be compared with a satellite digital decoder/receiver as we discussed regarding DBS television. The major difference is that a set-top box is designed for coaxial cable.

Access rights and security levels are set on a removable smart media card that allows users of the CATV digital services access to programming that has been paid for. The security and decryption levels are based upon complex algorithms that prevent unauthorized access.

CONCLUSIONS

This chapter emphasized video transmission systems such as cable (guided), airwave (unguided), and satellite systems (unguided). Television roots and early video transmission technologies were examined. Video devices such as the television receiver and television projector technologies were scrutinized. Video storage devices, such as the VCR, PVT, and DVD, were examined. We now understand that tremendous amounts of bandwidth are saved when digital video is used instead of analog video.

Let's switch gears a bit and take another look at the local exchange carrier, only this time we'll focus on high-speed data services.

5 High-Speed Data Services

Because of the Internet and our desire to interconnect with many different networks, broadband, high bandwidth services are here to stay. How to get access from the home to high-speed networks in a cost-effective manner is the issue here. If we cannot get high-speed access at a reasonable cost, then we must rely on "legacy" modems with a maximum speed of 56 kbps. While 56 kbps may have seemed "lightning fast" several years ago, it is quickly becoming outdated for today's graphics- and video-hungry users.

Modems can convert digital information into the analog public switched telephone network (PSTN) world and connect to a network. However, we must consider a network that will accommodate digital high-speed data and not retrofit this high-speed data into the existing analog voice network. This has been the major stumbling block of high-speed network access from the home thus far.

Much has already been said in our society about the Internet. There will be debates regarding Internet content for many years to come. Let's take a closer look at the technology and networking required for Internet access.

The first section covers packet switching and Internet access. The second section takes a look at TCP/IP, gateways, routers and IP addresses. The third section discusses digital subscriber lines (DSL). The fourth section covers cable modems services, and the last section, covers Internet satellite services.

Get ready for the information superhighway and the need for speed!

Brooom Brooom Brooom......

PACKET SWITCHING AND INTERNET ACCESS........

In many respects we are in a transition phase of telecommunications network technology. We are slowly moving away from traditional services, such as voice and low-speed data, toward high-speed data services. To provide these services, however, the network must contain the appropriate equipment and media that will support high-speed connectivity.

Packet Switching

The best technical solution to building high-speed data networks for a small business is to build an entirely new system with the latest technology. This new network would be designed to handle local data access, Internet connectivity, and email. Voice services would be available through a circuit-switched network similar to the PSTN.

Let's review and consider what happens in a circuit-switched network. Remember our telephone call in Chapter 3? The circuit is established at the time of call setup, and the circuit remains intact for the life of the call. If a caller decides to put the telephone call on hold, the circuit remains intact and the call remains established. A telephone call on hold still requires a connection and bandwidth until the call is dropped.

For now, let's examine a data circuit and try another solution.

We already know that we can send a digital data stream of 1's and 0's through the network. In a circuit-switched scenario, the circuit is "Always ON" until the call is dropped. Instead of an "Always ON" approach, let's try to divide the digital data stream into groups. This would be the equivalent of "packetizing" the data. In other words, we split the data into chunks or packets. Each packet would contain a destination address, a starting flag, and an ending flag. Packets would be variable-length units, each having a minimum and a maximum value. Each packet or chunk of digital data is sent from the origination point to the termination point and reassembled before conversion to analog (Figure 5.1).

In Figure 5.2, the packets containing digitized voice information are sent to the destination location by the fastest route. Packets 1 and 2 travel a more direct route; packets 3 and 4 travel a different route because of congestion on the direct route. Therefore, a problem exists whereby the packets are delivered out of sequence. The delivered packets must be reassembled in their correct order before conversion from digital to their original format. This reassembly must occur quickly or the information display will be delayed (Figure 5.2).

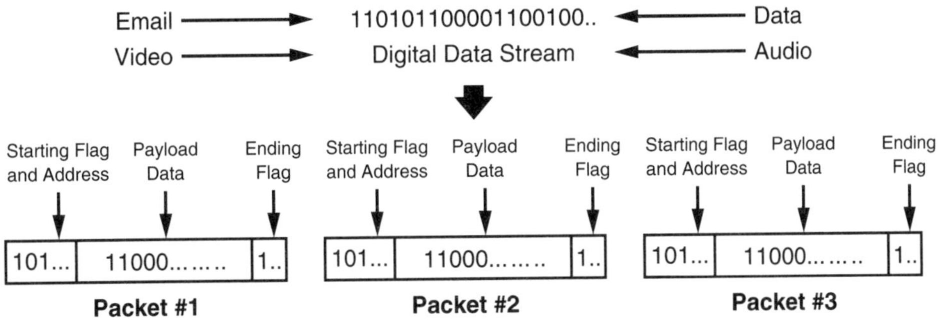

Figure 5.1
"Packetizing."

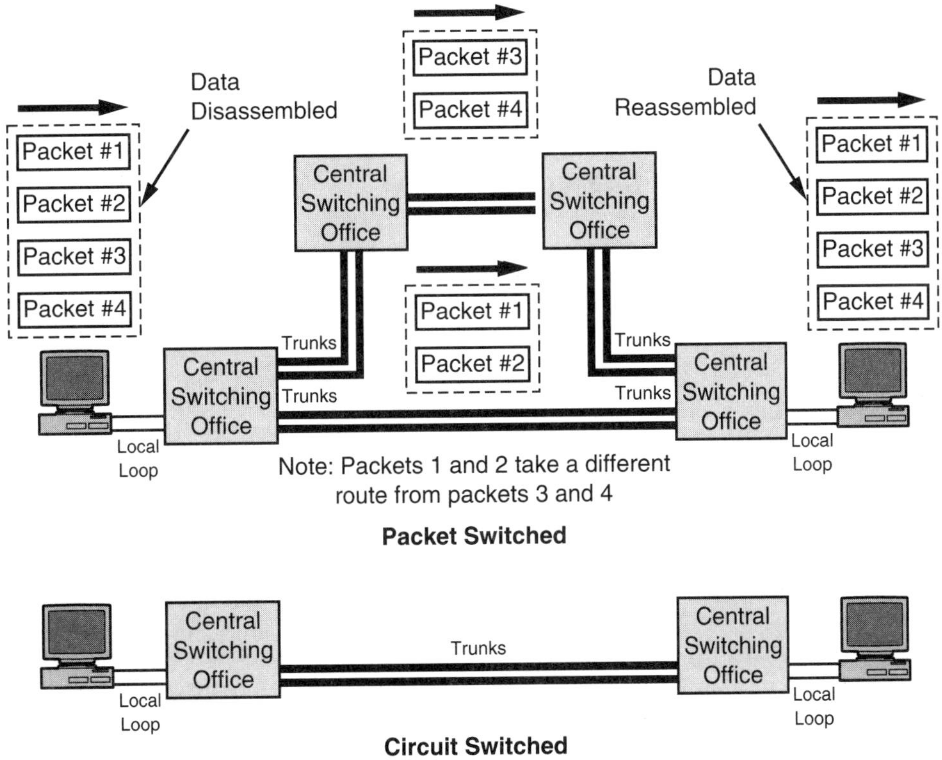

Figure 5.2
Packet switching vs. circuit switching.

OK, so you may be wondering why packet switching is necessary. Certainly, circuit switching does a fine job with voice transmission and is an established mature technology that has performed well for over 75 years. Actually, no one even talked about replacing circuit-switched services with packet-switched services until the Internet became universally accepted in the mid-1990s.

Basically, packet-switched services provide a more efficient telecommunications network. For example, if we were engaged in a telephone conversation and the conversation suddenly paused, there would be no information to transmit: just silence. In a packetized network, packets would not be sent during silent periods. Studies have shown that approximately 20% of voice conversation is silence (including pauses).

We will examine Voice over Internet Protocol (VOIP) solutions later in this chapter.

Let's reexamine packet switching on the Internet:

- Packets may not arrive at the destination address in the same order.
- Packets may never get delivered or get delivered to the wrong address.
- Packets may arrive corrupted or severely damaged.

Packets from PSTN Trunks

Now, take away voice services and add the Internet. Internet data is very bursty. There can be long exchanges of data and long periods of silence. On a circuit-switched network, bandwidth is used even when there is silence or no exchange of information. On a packet-switched network, however, packets are sent when information is requested and not sent when there is no information to be sent.

In other words, when bursty packets are sent on an intermittently used Internet connection through the PSTN, circuit-switched resources are tied up considerably. The overutilization of telephone company resources can be a problem. The trunking and overall capacity of a central office can be compromised. For example, the average call time before the Internet (1995) was 3 minutes. Presently, using Internet data through the PSTN, average call times have jumped to 60 minutes or more per call.

Furthermore, the lack of Internet access charges from the LECs assessed to Internet service providers has led to low connectivity fees of $20 or less to provide service.

Because of the problems with increased Internet data, switch blocking has occurred. You will recall that call blocking is the result of unavailable trunks and that switches are limited to the amount of connections they can perform.

LECs have been required to make huge investments to increase central office trunking and switching capacity (much to their dismay) because of dial-up data services through the PSTN.

Packets from PSTN Local Loops

The trouble with local loop access today is the 3 kHz bandwidth limits provided by analog services. The local loop was never redesigned even when digitized fiber trunks became available in the mid-1980s. It wasn't considered necessary to modify the local loop because the PSTN is still an analog voice service and the local loops could still support the technology.

Loading coils in the local loop are used to counteract resistance or attenuation of the signal. These coils also improve the analog frequency response of the voice signal by improving high-end voice frequency quality. Both loading coils and bridge taps cannot be used for high-speed digital services.

The presence of loading coils, bridged taps, and mixed wire gauges impose a nightmarish effort on the redesign the local loop before any attempt to create a digital path. Furthermore, LECs have not kept records on loading coils, bridge taps, and mixed wire gauges to determine what or where this equipment is.

The problem is that packet data services such as the Internet and email are exploding while voice services are decreasing or remaining flat. The LECs are in a position where they need to provide next-generation services to the home without digging up the streets at a considerably higher cost.

We focus on the bandwidth available for local loop digital services later in this chapter.

Internet Service Providers

First of all, the Internet is a packet-switched network. Circuit switching is not involved here. Packetized data is sent from an originating Internet service provider (ISP) to a destination ISP. We already know that packets can take different routes to arrive at the final destination.

Second, the Internet architecture is truly a World Wide Web with the network becoming a network of networks. Because of the common architecture and protocols shared by all, the Internet is a common area for data distribution. Because of this design, the Internet can only proliferate.

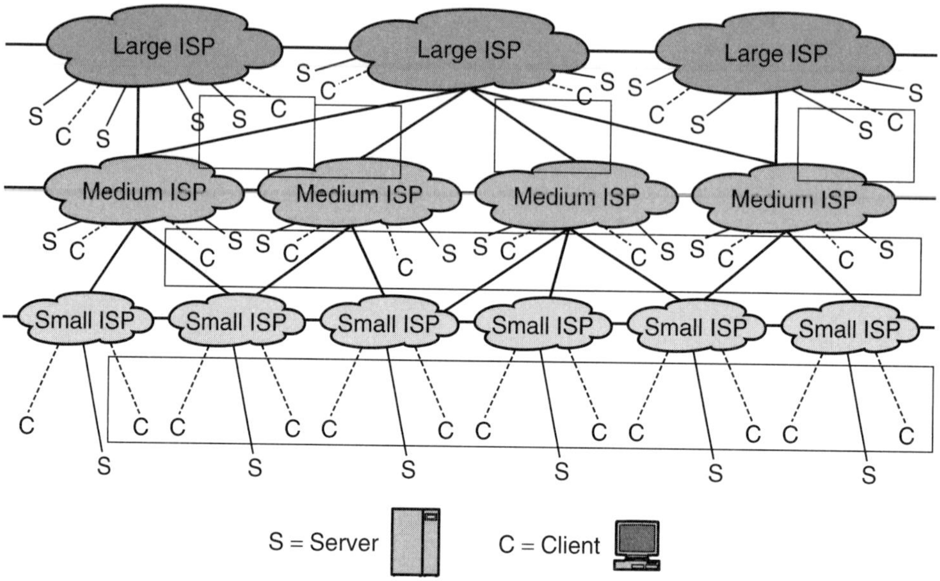

Figure 5.3
The Internet: World Wide Web.

The ISP is our first level of access to the packet-switched data network (Figure 5.3).

ISPs supply connectivity to other ISPs as well as access for local users such as ourselves. Some of the largest ISPs such as UUNET Inc. (an MCI Worldcom Company) and Cable and Wireless provide high-speed trunks specifically for interconnection of packet-switching services.

For example, let's say we want to access the XYZ Company (fictitious) Web site and they utilize UUNet Inc. as their ISP. We are using a PC with a 56 kbps dial-up modem and America Online (AOL) is our service provider. This is typical client-server technology, where the client is our PC and the server is at the XYZ Company.

Because of our increased desire to access the Internet for graphics, sound bites, and text, etc., the limitations of the 3 kHz analog channel have proved to be inadequate. Aside from the social problems of the Internet, many of the Internet's problems are associated with connectivity issues and the public's dissatisfaction with access speed and bandwidth.

A meeting place or exchange of data is needed to access UUNET or the XYZ Company's Web site from our America Online (AOL) ISP. This is known as peering. Peering is an agreed-upon data exchange between two ISPs (Figure 5.4).

TIMETABLE OF HISTORIC INTERNET EVENTS

1969	Len Kleinrock's lab, at the University of California (under contract with the Department of Defense), demonstrates the first packet-switching configuration between two mid-sized computers.
1970	Four university sites are set up with packet-switching configurations. This configuration is later dubbed "ARPANET."
1972	Ray Tomlinson introduces the concept of email on ARPANET.
1974	TCP/IP is developed containing file transfer, email, and remote access features.
	America Online adds dial-up access to its bulletin board system.
1989	ARPANET (Government) is changed to Internet and is now available to the general public. The Internet is later dubbed the World Wide Web.
1990	Internet is still completely text driven. UNIX commands are used.
1993	Mosaic browser is first used, providing a point-and-click graphical interface.
1995	Netscape browser becomes available as a free PC download.
1996	Microsoft browser, called Internet Explorer, becomes available to compete with Netscape's browser.

INTERNET PROTOCOLS AND CONNECTIVITY ....

Transport Control Protocol/Internet Protocol (TCP/IP) is a suite of protocols developed in the 1970s for the Department of Defense's ARPANET network, which later became the Internet. Protocols such as TCP/IP are important standards that have

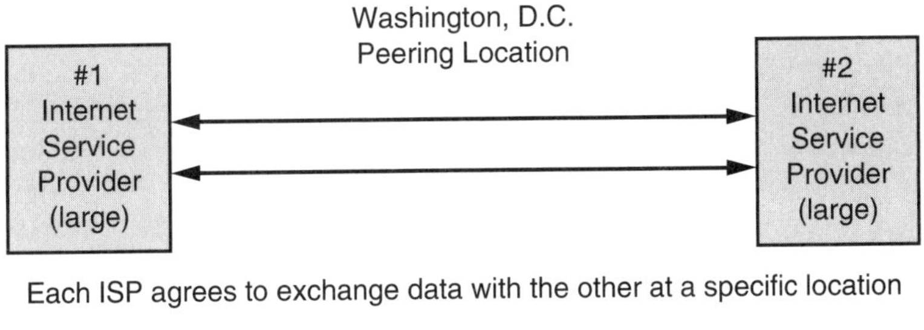

Figure 5.4
Internet service provider peering.

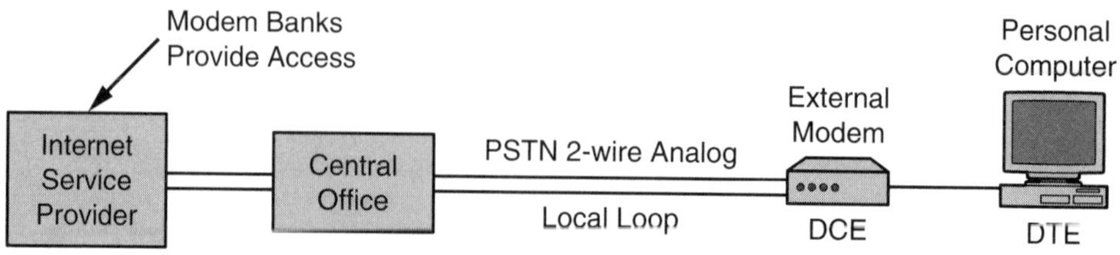

Figure 5.5
Internet dial-up access.

evolved over the years and are supported universally. The design of TCP/IP is for the Internet; however, these protocols can also be used for internal network configurations.

The most common method of access to the Internet is with dial-up modems. However, dial-up access is on the decrease as new and faster methods of Internet access become more affordable and available (Figure 5.5).

Before we discuss TCP/IP, let's determine what protocols are. Protocols are agreements between devices to provide communications compatibility between end-terminals. Simply stated, a protocol is the language that a network speaks. For example, humans choose a language such as English to speak and communicate with. Both humans understand each other. The Internet has a common language also. This language is TCP/IP. Let's investigate TCP/IP further.

TCP/IP Architecture

TCP/IP has a three-level hierarchy. The lowest level is the subnetwork; the next highest level is called a domain. The highest level of TCP/IP is the enterprise network connecting the highest domains together.

The IP portion of TCP/IP is a connectionless protocol. A connectionless protocol says that packets containing addresses are sent into the network without a defined transmission path. The TCP portion of TCP/IP contains error checking and acknowledgement features and is a connection-oriented protocol. Connection-oriented protocol says that a virtual path is created for data transmission from origination to destination.

Using the TCP/IP architecture model, we can trace a typical scenario for the File Transfer Protocol (FTP).

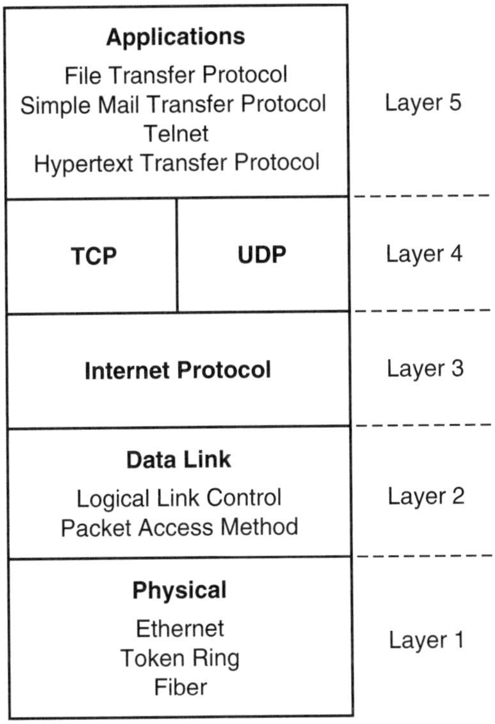

Figure 5.6
TCP/IP architecture model.

The applications layer transfers the necessary data to the TCP layer, which adds an address header. TCP then passes the data to the IP layer, which breaks up the data into datagrams and passes the data to the physical layer. A datagram is an individual packet sent through the network without any regard for other packets. The datagram then goes to the physical layer, which transmits the datagram to its final destination address (Figure 5.6).

Remember that in a packetized network using TCP/IP, datagrams are sent to their final destinations by the shortest path first. The datagrams may or may not arrive at their final destination in order. Each datagram arriving at its destination uses TCP, which ensures reliability and integrity in the network. Error checking is accomplished when the destination terminal sends an acknowledgement to the originating terminal that the datagrams were received in good shape.

IP Addresses

An IP address is composed of 4 segments separated by decimals, with each segment containing up to 255 addresses. This IP addressing scheme is known as IPV4. An example of an IPV4 address is 190.12.4.1. The first octet number contains the IP class. Address classes can be A, B, C, or D (Table 5.1).

Table 5.1 Addresses Classes and Ranges of Internet Protocol

Address Class	Address Allocation	Minimum	Maximum	First Digit (Binary)
A	host	1	126	1
B	14 bits host /16 bits network	128	191	10
C	8 bits host /21 bits network	192	223	100
D	Multicast	224	239	1110
E	(experimental)	240	247	11110

IP addresses are in short supply, especially at host levels. An expanded version of IP called IPV6 has been approved. We discuss more about IPV6 in Chapter 10.

The designers of TCP/IP knew that IP address numbers would be difficult to remember so they created a naming convention. Each user is part of a domain which is the name of the Internet service provider. The domain name is joined to a suffix and separated by a period. For example, an address could be *http://www.att.com.* Recently, there have been several attempts to standardize on suffixes such as .com, .net, .edu, etc., without much success.

The names are transmitted across the Internet to the domain name, which must then be correlated with an IP number address. This translation is handled by the Domain Name System (DNS).

In larger networks, IP address assignment for individual computer workstations can be somewhat of a nightmare. Administration of static IP address are further complicated when laptops are used since the IP subnet can change as the laptop is moved. A solution to this dilemma is to use Dynamic Host Configuration Protocol (DHCP). DHCP maintains a pool of IP addresses. DHCP saves administration time and conserves IP addresses that may not always be available.

Routers and the Internet

Routers are used in packet-switching networks. Routers are intelligent devices that determine the best route for a packet. A router knows all the available paths and can determine the shortest delivery path and least expensive delivery path, and it can look to avoid congestion problems. If a packet can reach its destination directly from a router, the address will be in the router's routing tables. However, in most cases, a packet will travel through several routers before reaching its final destination.

Router performance is based upon the packet-forwarding rate. This rate specifies the number of packets per second. Routers may need to split variable-length packets even further. For example, if a packet-switched network can handle a maximum packet size of 1,000 bits, the router would fragment the packet to make it fit the network.

Routers can use a variety of protocols. Our concern here, however, is with TCP/IP. These devices will periodically send a short test message to see who its neighbors are and determine the number of hops to each destination (each router is a hop). In Figure 5.7, we can see that route #3 is the best circuit path to route a packet to its final destination. Route #3 via Dallas may not be the shortest circuit path. However, this path has no congestion and that makes it more desirable for routing a packet.

Routers can also be configured as a firewall. A firewall can block packets between unauthorized networks.

Gateways and the Internet

Gateways select the best route to a destination address. Today, the function of a gateway device is performed by a router.

Gateways can be one of two types. Some gateways contain extensive network information (core gateway) and others contain limited network information (noncore gateway). Packets are sent through core gateways in an attempt to get as close as possible to a destination address. The final gateway may be a noncore gateway that is directly connected to its ultimate destination.

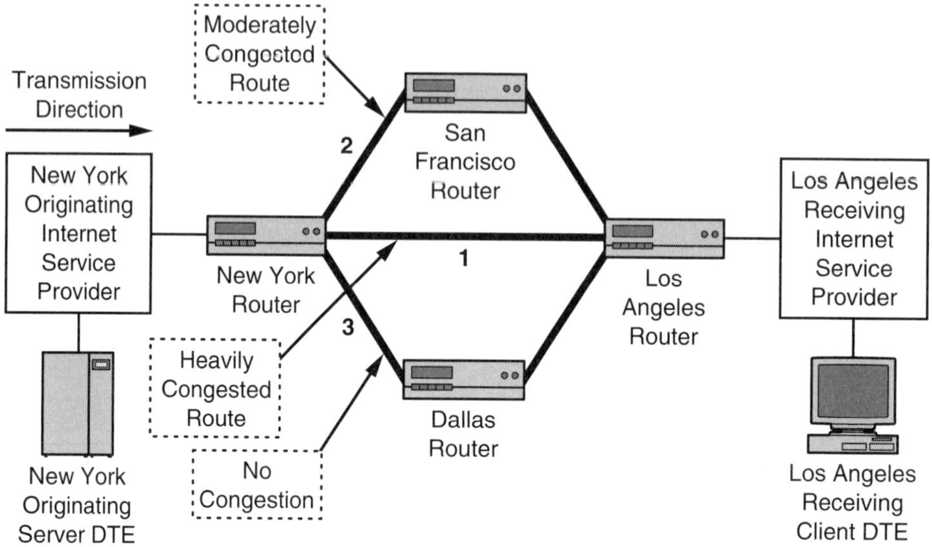

Figure 5.7
Routers and the Internet.

Firewalls and the Internet

Firewalls are increasingly becoming a necessary device that protects data from security threats from Internet hackers and harmful viruses. The function of a firewall is to filter packets. The firewall inspects the incoming packet and determines if a block is necessary. For example, a firewall can be programmed to block all packets except from those of a coworker. The coworker's TCP/IP address would tell the firewall that these packets should be allowed to reach the client's DTE.

Unfortunately, packet filtration is not always effective. An experienced hacker can use a technique called spoofing. This technique is the falsification of an origination address in the packet by a hacker to gain access to server resources.

BROADBAND SERVICES WITH LOCAL LOOPS.........

The term "broadband" describes a high-speed data transmission scheme where multiple transmissions share the same path. Many of the telecommunications services we've already talked about are broadband services. However, broadband services to the home are relatively new.

Table 5.2 Book (250 Page) Download Times

Bandwidth	Time	Speed
2400 bps	90 minutes @	.06 pages/sec
33.6 kbps (Standard Analog)	06 minutes @	.87 pages/sec
128 kbps (ISDN-BRI)	1.7 minutes @	3.34 pages/sec
1.5 Mbps (T-1)	7.7 seconds @	.10 books/sec
6 Mbps	1.9 seconds @	.50 books/sec
45 Mbps (T-3)	.25 seconds @	~4 books/sec
155 Mbps	.07 seconds @	~14 books/sec

The current interest in broadband services is due almost exclusively to the Internet and residential high-speed access. Today's data-hungry applications require more bandwidth to run better and more efficiently. However, there can be many bottlenecks in a telecommunications network. For example, a major IP trunk belonging to an IP carrier can become congested. This congestion may delay or impede any data supplied by the carrier, making some addresses available while others are unavailable. The client would not have any knowledge of network congestion and could assume (although incorrectly) that the local ISP was at fault.

Whether we choose digital subscriber line (DSL), cable modem, or Internet satellite for broadband access from the home, it is important that we understand the basics of each technology, the options that are available, and the cost for each service (Table 5.2).

Digital Subscriber Line Services

Availability is the key to DSL services. First, you must investigate the availability of service, the costs of service, and the way in which the service can be integrated into the home. Some customers have committed a portion of their monthly expenses to xDSL connection costs, while others have given up and have relied on their standard 56-kbps modems using the PSTN.

DSL is a "quick and dirty" way of providing broadband services to the home. Until the Internet became popular, there was no real concern to modify or "digitize" anything in the network for local access. The analog 3 kHz local loop proved to be acceptable for voice. However, bandwidth requirements were getting greater and greater. Digging up the streets was not an option (too expensive) for adding new lines. Moreover, wireless broadband services were too costly.

One of the few cost-effective solutions was to reengineer local loops. There proved to be several real world problems with local loop redesign:

- Approximately 20% have loading coils (no records were kept on where the coils were installed).
- Bridgc taps and mixed wire gauges were installed.
- Average local loops had 2 dozen splices. These could be corroded, adding attenuation.
- Distances between residences and central offices averaged over 11,000 feet.
- Line extenders (amplifiers) were sometimes used. This use extended the local loop to 25,000 feet or more.

The first DSL was BRI-ISDN (BRI-ISDN was not called DSL at the time). Many of the features of DSL were made possible through experience with BRI-ISDN. Remember that BRI-ISDN provides one 128-Kbit channel (bonded) and one signaling channel (16 Kbits). With BRI-ISDN, the local loop was digitized and the analog 3 kHz limitation was removed.

All DSLs terminate into a central office digital subscriber line access module (DSLAM). The DSLAM aggregates all of the data from each of the DSLs onto one large pipe (usually fiber) for access to the Internet and the packet-switched network.

Remember that DSLs are not switched services. The services are "Always ON" and available for IP data. Table 5.3 lists the types of DSL services.

Table 5.3 The DSL Family of Services

Service	Bandwidth	Mode	Type
HDSL	1.544	Symmetrical	Two pairs (copper)
ADSL	1.1 to 8 Mbps (down) 16–640 kbps (up)	Asymmetrical	One pair (copper)
CDSL	up to 1.1 Mbps (down) 16–128 kbps (up)	Asymmetrical	One pair (rate adaptable) (copper)
IDSL	(same as BRI-ISDN)	Symmetrical	One pair (copper)
VDSL	13–52 Mbps (down) 1.5–6 Mbps (up)	Asymmetrical	Fiber needed

TIMETABLE OF HISTORIC DSL EVENTS

1989	DSL is first developed to compete with cable for video services. DSL gives the telephone company the ability to provide future video-on-demand services.
1990	HDSL is developed by Bellcore, reducing the need for repeaters.
1993	ADSL is developed. It contains different uplink and downlink speeds.
1995	Internet browsers become available.
1997	ADSL implementation first begins. It provides low-cost, high-speed Internet access.

Before we begin to analyze these services in more detail, let's look at competitive DSL providers vs. traditional LEC providers.

Competitive Access Providers and Colocation

In an effort to foster competition in the telecommunications local access arena, the FCC and the Justice Department in 1992 allowed small companies to compete with local exchange carriers (LECs). These competitive access providers, or CAPs, were primarily focused at the business user but would serve residential customers also. Examples of CAPs in the DSL market are Covad Communications Company and Rhythms Net Connections Inc.

The problem for CAPs is that they still need to interface with the 2-wire local loop that is owned and maintained by the LEC.

The best and least costly solution is to allow CAPs to collocate their equipment in the central office. The LEC provides an area in the central office building (usually within a caged area or locked room). The LEC charges a reasonable fee for access to residential local lines and a reasonable fee for space in the building. Without colocation, CAPs could not survive because the costs to provide service would be prohibitive (Figure 5.8).

The LECs still have an advantage over CAPs providing DSL services. CAPs still must coordinate their efforts between themselves and the LECs who will provide access to the local loop and ultimately the customer. The end result has been slower installation and greater frustration for customers using CAP services. The breakdown is shown in Figure 5.9.

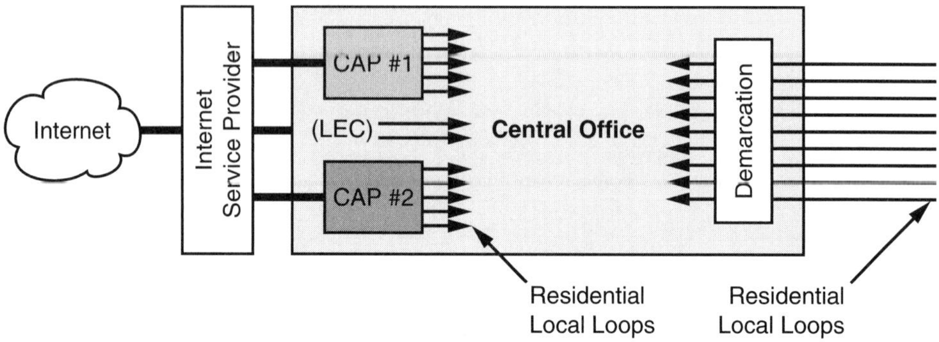

Figure 5.8
Competitive access providers and local Internet access.

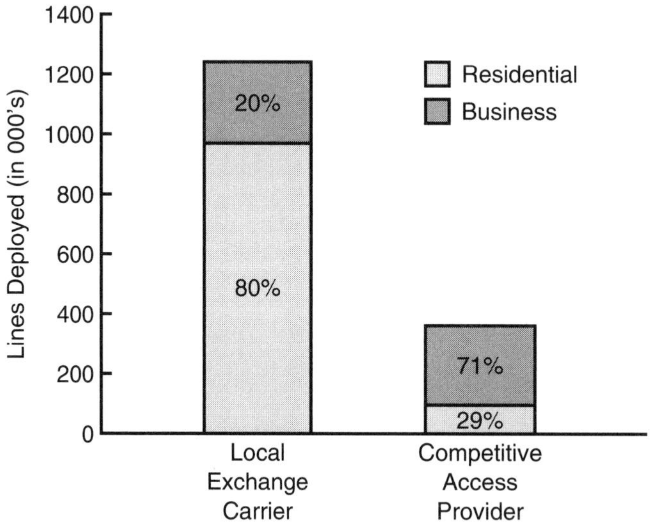

Figure 5.9
Breakdown of DSL residential and business subscribers.
After TeleChoice, Inc./Turnstone, Inc.

Asymmetrical Digital Subscriber Line

Asymmetrical digital subscriber line (ADSL) has the potential to supply small business and residential users with all types of broadband services, not just Internet capability. Interactive video services and high-resolution graphics are some of the capabilities of ADSL.

The word *asymmetrical* denotes different speeds in different directions. Up to 384 kbps are available in the upstream direction, and 768 kbps are generally available in the downstream direction.

ADSL contains both upstream (transmit) and downstream (receive) operations in a full-duplex (bidirectional) configuration on a single copper wire pair. Remember that plain old telephone service (POTS) uses a single circuit or one copper wire pair for standard voice operation. With ADSL, the same pair of wires is used; however, the frequencies must be split, or echo cancellation must be used. In either case, frequency division multiplexing (FDM) is used.

Echo control is generally required whenever the same frequency range is used for both upstream and downstream operation. Echo generally occurs in full-duplex circuits because some of the signal is reflected back to the sender, creating interference and, ultimately, errors in the data. Echo cancellers electronically subtract the signal sent from the signal received.

Another method to eliminate echo is to segregate the frequencies. A frequency range of approximately 25 kHz to 200 kHz is available for upstream data, and a larger frequency range of approximately 200 kHz to 1.1 MHz is available for downstream data (Figure 5.10).

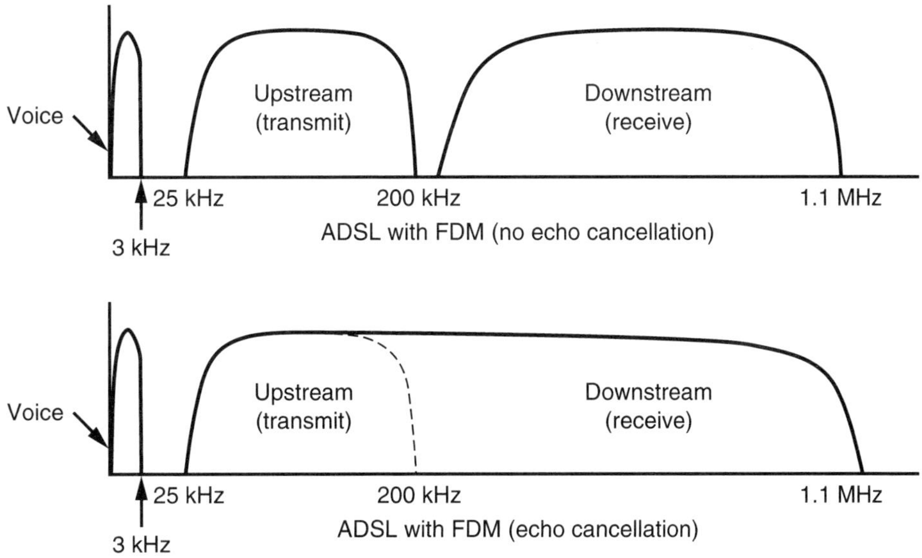

Figure 5.10
ADSL bandwidth.

Also notice that the ADSL bandwidth chart contains frequencies assigned to traditional analog voice (3 kHz). A signal splitter used at the residence or small office allows both voice and data to be shared on the same line.

The assumption is that when downloading data, more bandwidth is needed for graphics, images, etc. In many cases, the only data sent up to the network is the TCP/IP address or Uniform Resource Locator (URL) address (e.g., *www.att.com*). Therefore, greater speeds are needed for download vs. upload, making the line asymmetrical.

Using one copper pair is an advantage; however, the need to support full-duplex operation on one line limits distance. In general, the maximum distance for ADSL is 18 kft (3.4 miles) from the central office. Of course, many customers do not fall into this category and ADSL service cannot be provided.

ADSL products use either one of the following coding techniques:

- Carrierless amplitude phase/quadrature amplitude modulation (CAP/QAM)
- Discrete multitone technology (DMT)

Line codes are needed to determine how digital signals are processed. CAP/QAM uses the entire local loop bandwidth (with the exception of the voice frequencies) to process the digital data. Instead of traditional FDM, CAP/QAM uses a combination of FDM and phase modulation.

DMT uses approximately 300 subchannels to process digital information. In most cases, the lower subchannels are reserved for traditional analog voice. There are 32 upstream subchannels, usually starting at subchannel #7, and 250 downstream subchannels. Each of the subchannels uses a 4.3 kHz bandwidth.

When an ADSL device that has DMT signal coding is turned on, each of the subchannels goes through a testing process. It is a modem handshaking procedure × 300! The noise, frequency response, and attenuation levels for each channel are monitored for performance.

The speed of an individual subchannel can actually vary depending upon the results of the handshaking/testing phase. DMT speeds can decrement in 32-Kbit steps. In other words, an ADSL DMT device can operate at 768 kbps or 736 kbps, depending upon operational and environmental conditions affecting local loop performance (Figure 5.11).

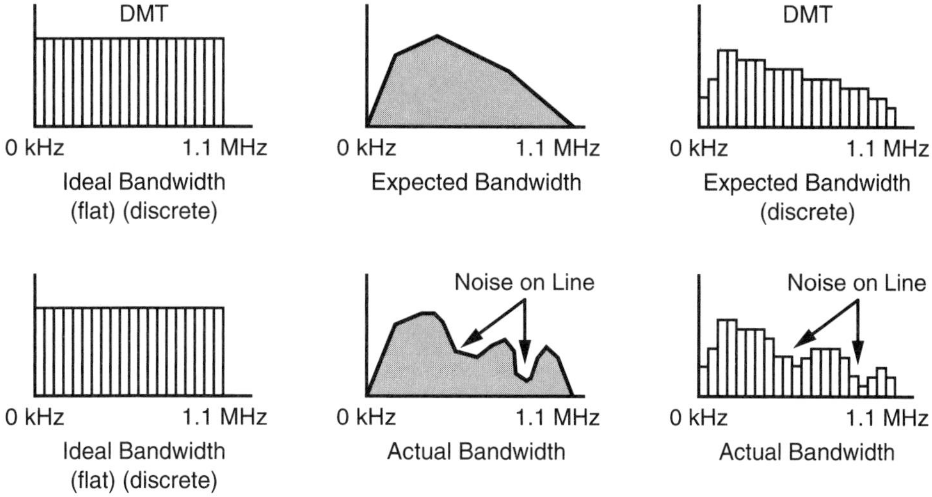

Figure 5.11
ADSL and DMT.

Ideally, we would like to see the same local loop specifications for each of the subchannels. Noise characteristics, frequency response, etc., would be even across the frequency spectrum. In reality, age of copper wire, length of wire, solder joints, outside interference, mixed wire gauges, etc., affect the subchannels differently.

At the higher frequencies, the length of the wire is a factor as the higher frequencies become attenuated. At the lower end of the frequency spectrum, noise and crosstalk may become a greater factor. Nevertheless, ADSL is designed to handle any type of line condition up to its maximum 18 kft distance.

DMT has essentially replaced much of CAP/QAM for ADSL line coding. DMT's subchannel optimization is its strong point. In fact, because of DMT's flexibility in analyzing line conditions, ADSL is sometimes called rate-adaptive asymmetric digital subscriber line (RADSL).

High-Bit-Rate Digital Subscriber Line

T-1 services (1.544 Mbps) use two pairs of copper wires and repeaters or amplifiers (approximately every 6 kft) to boost the signal up to acceptable levels. Because of the repeaters required, T-1 pricing remains fairly expensive for residential or small business use. To minimize and reduce pricing, engineers at Bellcore developed the high-bit-rate digital subscriber line (HDSL) as a repeaterless T-1.

A big advantage of HDSL is that it gives the LEC the ability to "turn up" the service quickly and at reduced cost. Without repeaters, the HDSL signal can be transmitted up to 12 kft from the central office.

Consumer Digital Subscriber Line

We've already discussed ADSL voice services. Providing these services over DSL-configured local loops, however, requires a splitter at the residence or small office. If the ADSL splitter is not installed correctly, is missing, or is just not working correctly, problems can result. To alleviate problems, splitterless DSL, or consumer digital subscriber line (CDSL), was developed.

With CDSL, there is no need for a technician to install splitters at the time of installation or provide future on-premises technical support. Because there is no splitter at the customer's premises, speeds are held to 1.0 Mbps. The 18,000 ft distance from the central office is still the same for CDSL as it is for ADSL. In fact, ADSL and CDSL are so similar that CDSL is sometime referred to as ADSL-lite (Figure 5.12).

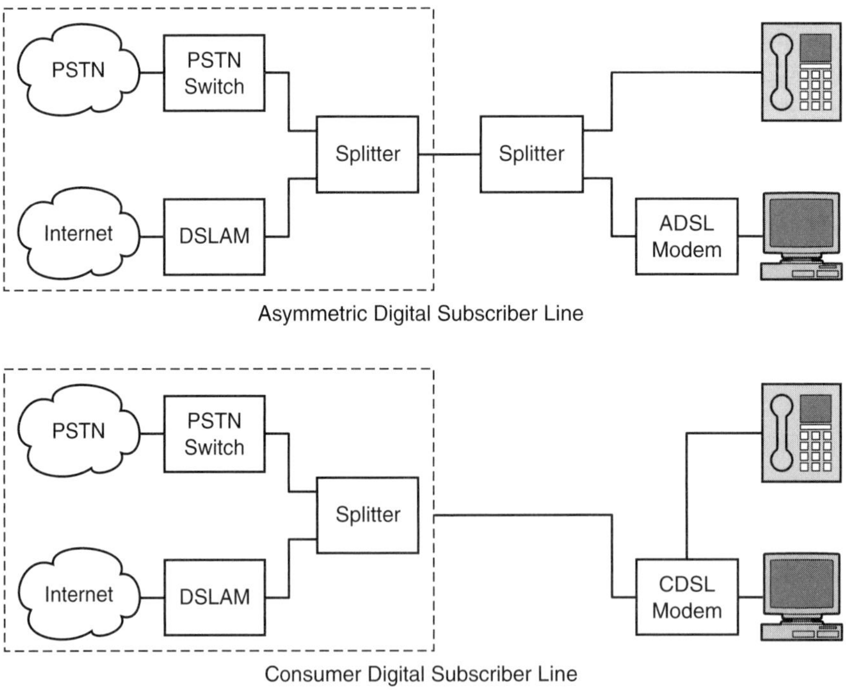

Figure 5.12
ADSL and CDSL configuration.

Very High Speed Digital Subscriber Line

Very high speed digital subscriber line (VDSL) is designed for optical local loops (not traditional copper pairs). Both the downstream and upstream bandwidth is much greater for this service than for the other DSL services already discussed.

It is not unusual for a new neighborhood or development to be served by fiber loops instead of traditional copper pairs. VDSL takes advantage of the benefits of fiber to provide downstream speeds between 12 Mbps and 52 Mbps. No small amount indeed!

Integrated Services Digital Subscriber Line

Integrated services digital subscriber line (IDSL) is similar to BRI-ISDN with one exception: it doesn't use the PSTN switch in the central office. Instead, it connects directly to the IP network. Since the PSTN switch is not connected here, voice services are not available. IDSL provides a downstream connection of 128 kbps, which is much slower than the other DSL services previously discussed. However, IDSL can be extended to 44 kft or more, providing DSL-like services to areas that are far away from the central office (Figure 5.13).

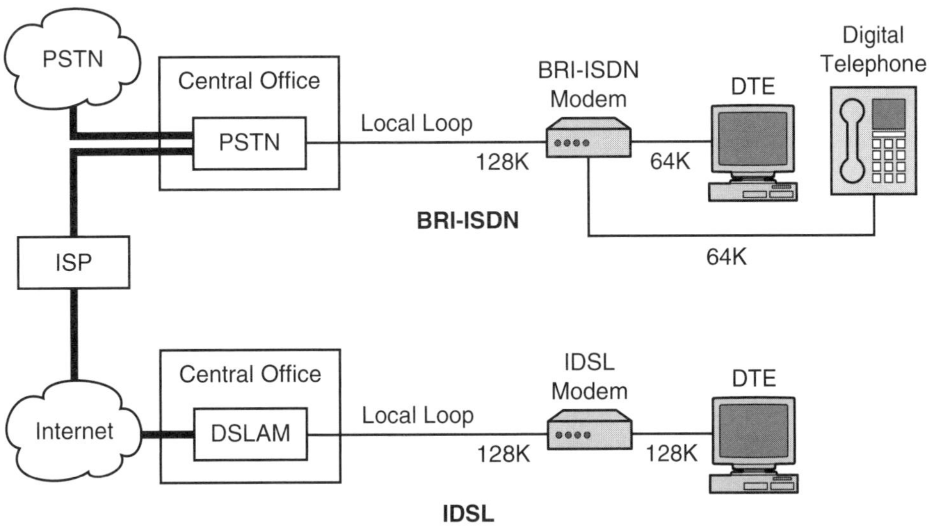

Figure 5.13
BRI-ISDN and IDSL comparisons.

DSL Issues

Problems with DSL implementation still exist today. Bridge taps and loading coils are some of the main problems. Splitters also contribute to installation delays (Table 5.4).

DSL is not the only available option providing high-speed Internet access. Let's explore some other options.

Table 5.4 DSL Issues*

Issue	Impact	Solution
Existing DSL test equipment cannot see through splitters in a DSL over POTS environment.	• Service providers rely upon engineering inventories to tell them whether a loop is qualified for DSL. • No real testing is done to verify the ability of a loop to support DSL, and improper determinations can result. • Delays in provisioning and customer frustration frequently result.	• A few providers, including at least one incumbent, are now placing DSL test equipment in front of the splitter. • For competitive providers, this is a solution only if there is sufficient room to colocate a splitter in their cage.
Splitters create need for additional wiring in the CO. With additional wiring, cross-connections are often mishandled, and when this occurs, providers have no direct way of knowing what the problem is.	• Because there is no means of remotely testing wiring in the central office, a provider has no way of checking whether connections have been done properly or done at all before service is supposed to be turned up. • When a problem does arise, a technician has a laundry list of items to check, including cross-connections. • Frequently leads to delays in provisioning and customer frustration.	• Wiring issues in the central office are dealt with largely through a process of elimination. • One incumbent provider is hardwiring binder groups of 100 pairs in an effort to cut down on labor and room for error, but there is no means of automatically verifying cross-connects.

* *Source:* TeleChoice, Inc./Turnstone, Inc.

BROADBAND SERVICES WITH COAXIAL CABLE......

In Chapter 1 we talked about different types of guided media. We examined twisted pair, coaxial cable, and fiber, and we determined that fiber offered virtually unlimited bandwidth. Let's not discard coaxial cable (coax). This media can transmit 10,000 simultaneous voice conversations, 500 6 MHz uncompressed television channels, and more. That is still a lot of bandwidth. The best news of all is that over 95% of residential homes and apartments in the United States are wired.

Logically, the cable companies decided to take some unused bandwidth from coax and provide consumers and small business access to the Internet. We'll discuss comparisons between all of today's broadband services later. For now, let's examine the cable modem and its network.

Cable Modem Services

A cable company offering Internet access will utilize a single television channel to provide Internet downstream services. This single channel occupies 6 MHz of bandwidth just like a normal uncompressed television signal. The big dilemma with cable modem service is that all data on the single channel is shared among several hundred users. This can be a security problem for both residential and business customers alike. More about the security aspects of cable modem technology in a later chapter.

The upstream data occupies 2 MHz of cable bandwidth, which is considerably smaller than the downstream side (6 MHz). Again, the concept is that less information is transmitted upstream than is received downstream (similar to ADSL).

The standard architecture for cable modem access is called data over cable service interface specifications (DOCSIS).

Two types of equipment are needed to establish service. The cable modem itself can be internal or external to the PC. If the local cable company has upgraded to digital MPEG2 services, a set-top box can be used to provide cable modem/Internet access.

The other important piece of equipment is the cable modem termination system (CMTS).

Cable Modem Termination System

CMTS provides a function similar to the DSLAM (see "Digital Subscriber Line Services" earlier in this chapter). The CMTS, located at the cable company's head end,

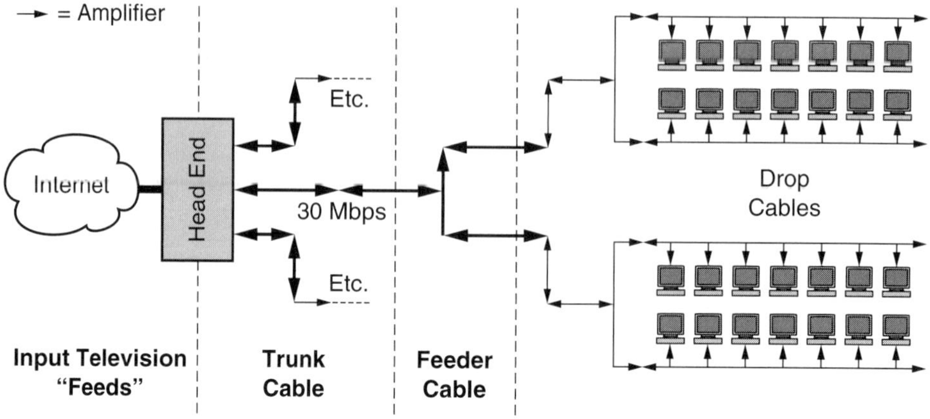

Figure 5.14
Cable modem networking (bidirectional).

aggregates upstream data from several hundred users into one large data pipe. This large data pipe is connected to the Internet.

A connection to the CMTS will allow up to 1,000 simultaneous users the ability to access up to 30 MHz of bandwidth. Remember, however, this is a shared connection with several hundred users simultaneously accessing the network. In general, cable network throughput will greatly improve when fewer of cable modem neighbors are accessing the Internet (Figure 5.14).

Figure 5.14 shows cable modem service as bidirectional. This is an important fact since we must be able to upload as well as download data. Unfortunately, cable television systems are designed for one-way simplex transmission only. The cable network amplifies television signals in one direction only: toward the television viewer.

Many cable companies have invested heavily in creating a bidirectional full-duplex network. In many cases, feeder cables must be upgraded to fiber to accommodate the incoming signals from the user's DTE.

You may be wondering what happens if the cable network is not set up for bidirectional, full-duplex transmission. Generally, in a simplex, one-way cable network, a standard modem is used to upload data and the cable is used to download data.

Sending and receiving data on two separate networks is not the preferred method of transmission. It is preferable to operate in a full-duplex mode through cable when possible. The cable companies want to provide you with Internet service as quickly as possible and will get you up and running by using a PSTN line for data upload wherever possible (Figure 5.15).

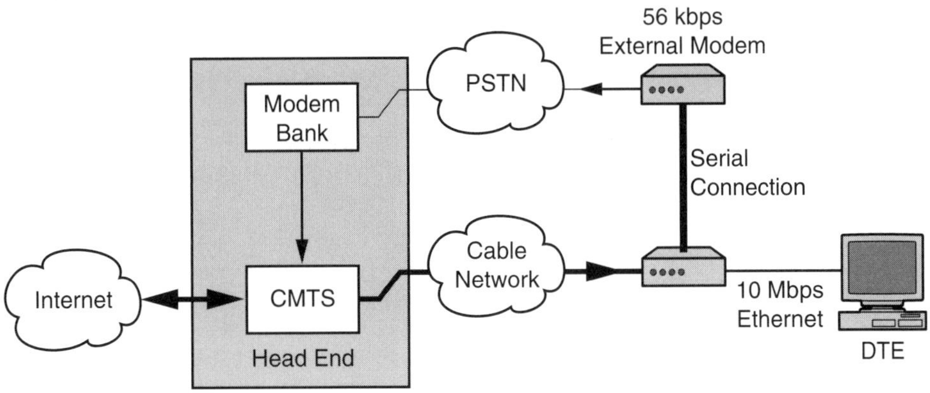

Figure 5.15
Cable modem operation.

It is important to understand that cable modem service is "Always ON" and that since it is a shared system it is vulnerable to attacks and hackers. My advice is to turn off your computer attached to a cable modem when it is not in use when at all possible!

We discuss security aspects of all broadband services in Chapter 9, but first we have one more broadband technology to discuss. Broadband services using DBS satellite transmission are now available.

BROADBAND SERVICES WITH DBS SATELLITE........

If you live more than 18 kft from the central office and if cable modem access has not been implemented in your area, you still have one other option: Internet satellite service. Internet satellite is great for users who want broadband access now but live in rural areas that have no existing broadband services.

Pegasus Express offers two-way Internet service under the brand name of DirectPC. Tachyon offers the same service in the United States, Mexico, and Western Europe.

An upgraded DBS satellite dish and receiver are needed to access the Internet satellite, providing 400 MHz bandwidth (downstream), which is less bandwidth than both cable modem and ADSL.

The Internet satellite receiver dish is set up just like a DBS dish. In fact, some DBS companies offer dual dishes that can provide digital television programming and

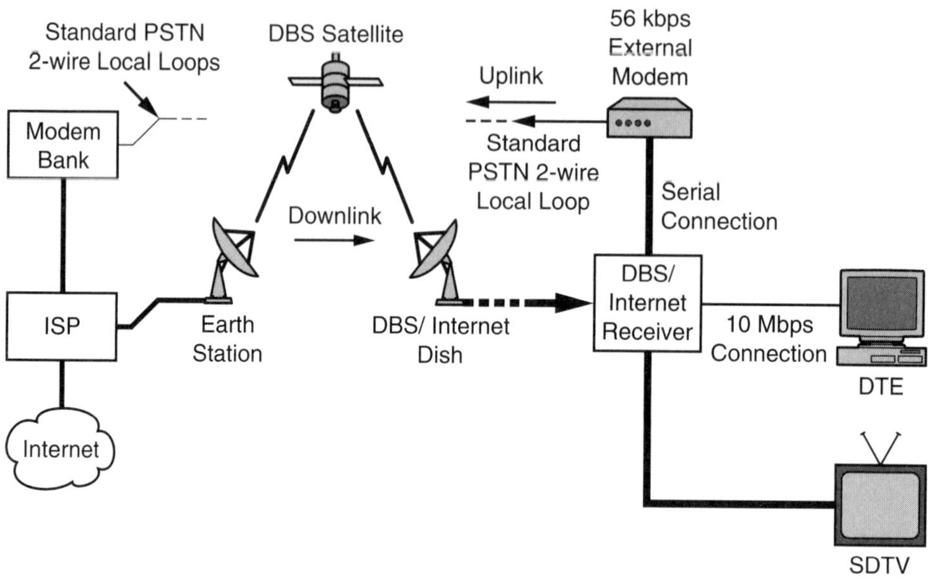

Figure 5.16
Internet satellite operation (simplex).

Internet access combined. Any DBS/Internet dish must be pointed directly at the satellite, which is in a geosynchronous position over the equator. Remember, heavy rain or other bad weather can affect any satellite transmission signals, including those of an Internet satellite.

A two-way Internet satellite uses IP multicasting technology. Multicasting provides service for several thousand simultaneous users. One-way service uses a standard modem for upstream data and the satellite connection for downstream data (Figure 5.16).

Remember that all Internet data is broadcast from the satellite, so this service is not a secure connection and should not be used when sensitive information is to be received.

CONCLUSIONS

This chapter has introduced broadband data services. The desire for high bandwidth at residential and small business levels has sparked several competitive DSL companies and has changed local exchange carrier responsibility from a PSTN local provider to a local provider of PSTN and high bandwidth services.

We now know that packet-switched services and TCP/IP are the future of telecommunications technology. They provide efficient worldwide digital transport of voice, video, and data services.

Broadband services are here now and provide the necessary access into the Internet at reasonable rates. Of course, today's broadband technology will be tomorrow's low-speed access. Do you know anyone using a 9.6-kbps modem today?

We now have a general understanding of telecommunications and networks; we know the basic requirements of analog and digital transmission, protocols, circuit-switched, packet-switched, and dedicated services.

One more thing before we begin the next chapter. I've supplied a reference chart that outlines the different services and speeds available for Internet access (see Figure 5.17).

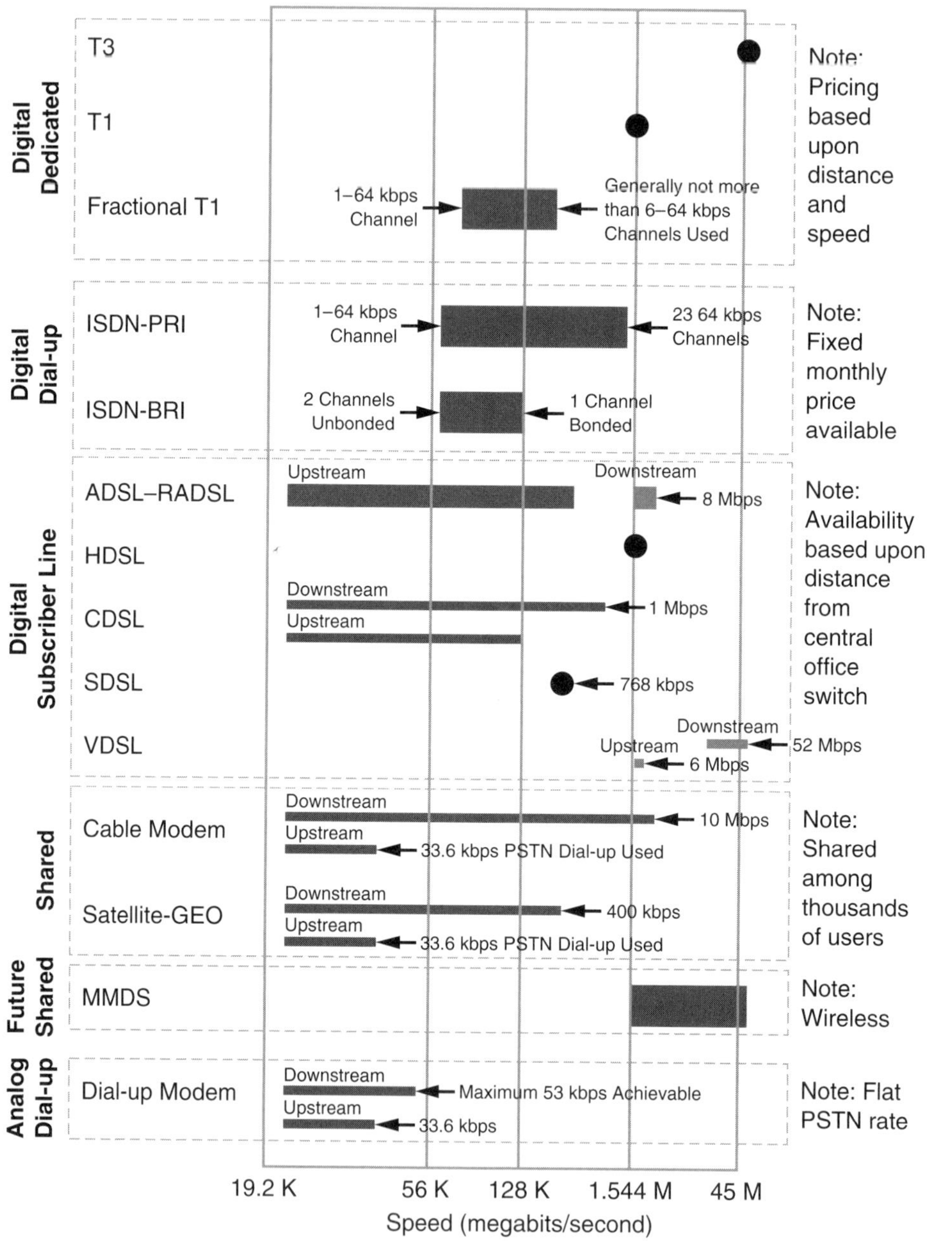

Figure 5.17
Internet access chart.

6 Virtual Private Networking

In Chapter 3, we discussed the concept of a private dedicated telecommunications network. Private networks are available 24 hours a day, 7 days a week. Private dedicated networks that are underutilized waste money (and bandwidth), and overutilized private dedicated networks can lead to data dropouts and errors.

Any alternatives? Well, a virtual private network can provide all the benefits of a 24x7 dedicated service plus managed features, consolidated billing, lower cost, and more.

In this chapter we also look at X.25 packet-switched networks, which were developed before the digital telecommunications revolution began. X.25 is still in use today, especially with financial and banking institutions.

Frame Relay networks, developed in the early 1990s, were designed for data traffic as businesses began migrating from voice to data. Frame Relay provides a cost-effective solution for business-oriented data networks.

This chapter virtually begins here.

VIRTUAL PRIVATE NETWORKS

A virtual private network (VPN) behaves much like a dedicated private line combining voice, video, and data on one customized network. A telecommunications carrier is usually responsible for configuring and managing VPN service. VPNs use "open network" architecture to transport packets. This is accomplished by the use of tunneling and encryption protocols over an IP network such as the Internet. IP VPN tunnels provide secure transport of data by means of the Internet backbone. VPNs can also be designed to connect with external networks such as PSTN and cellular networks.

Virtual Private Network Advantages

Many businesses, both large and small, look at VPNs as an alternative to dedicated private line services. VPNs provide private line functionality without the high costs associated with managing private line networks. Some advantages of VPNs are as follows:

- Consolidated billing—Video, voice, and data charges are invoiced on one bill.
- Cost efficiency—Minimal requirements for dedicated access circuits.
- Switched services access—PSTN, 800# dial access, remote calling card access, internal voice network access are available.
- Performance monitoring—A telecommunications carrier can increase network efficiency by monitoring performance, performing reroutes, or making network suggestions.
- Voice features—A telecommunications carrier can create caller groups, enabling or disabling voice calling features, etc.
- Dedicated "like" features—A telecommunications carrier can reallocate existing bandwidth as needed.
- Upgradable—Easy to upgrade and add new location or perform moves.

A VPN customer can view call detail and networking summary reports. These reports are used for traffic engineering studies and possible VPN reconfigurations. Customers can make suggestions, but the VPN is essentially the telecommunications carrier's responsibility. The purpose of VPN design is to provide one network for one company at reduced cost.

Virtual Private Network Architecture

Instead of a T-1 line configured for voice and a T-1 line configured for data, the architecture of VPN services provides for different services over one or several T-1 access lines. Numerous access arrangements are available from VPN service carriers. However, access lines must be provided from the customer's office to a carrier's network (Figure 6.1).

Having all services under one network provides substantial cost savings. A telecommunications carrier or vendor can perform traffic and cost analysis studies to determine if a VPN is economically feasible. An IP VPN can provide access at multiple sites around the world to create a flexible, easily expandable solution for any business application.

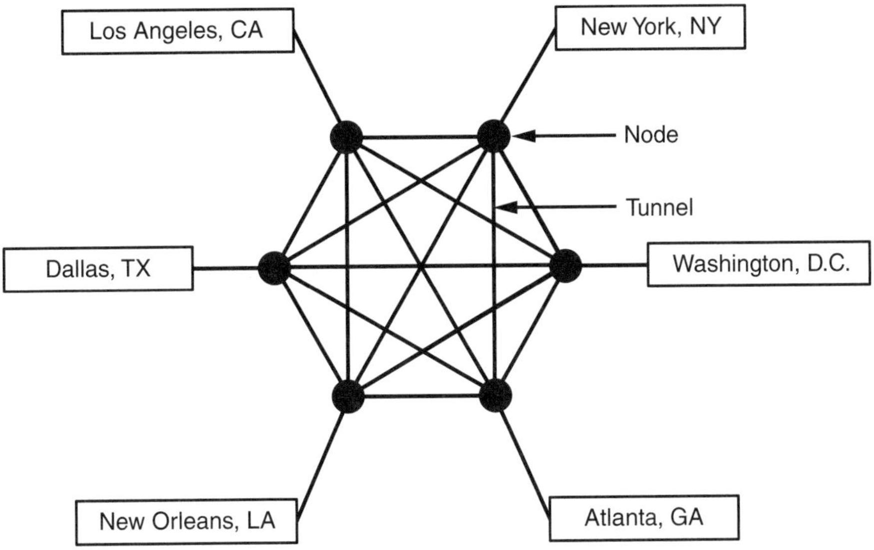

Figure 6.1
Virtual private network.

X.25 SERVICES

X.25 is another example of a virtual private network service. X.25 uses a packet-switched network with standards set by the International Telecommunications Union (ITU) in the early 1970s. X.25 was designed to support speeds up to 64 Kbps. At that time, backbone networks consisted of analog copper lines and electromechanical switches. To deal with this relatively hostile environment, engineers developed X.25 with error detection and correction in mind.

X.25's error correction capabilities were needed for poorly performing analog lines. Any node on the network could request a retransmission of the offending data from the originating node.

You may think that X.25 networks would not be needed in an environment where digital circuits and fiber backbone are prevalent. Actually, X.25 is extremely useful when reliability and not speed is the utmost concern. Credit card sales, financial information, and point-of-sale transaction processing are some examples of X.25 transmission.

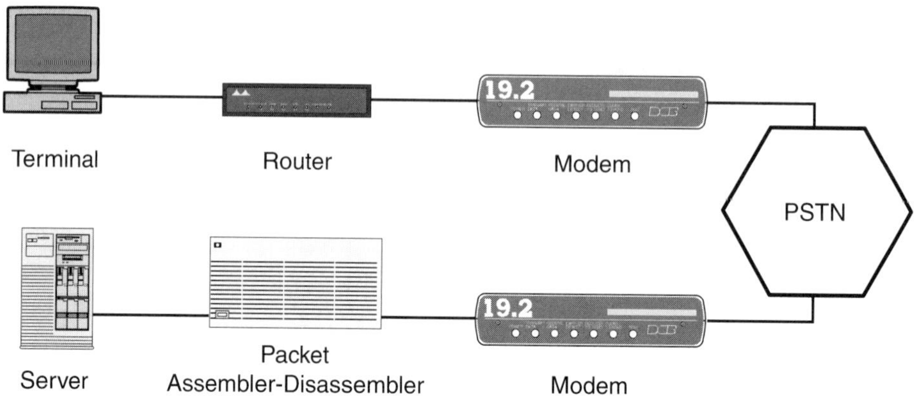

Figure 6.2
X.25 service configuration.

X.25 Network Architecture

X.25 architecture specifies the following:

- Datagram—Simple packet structure that contains enough information for the packet to be routed to a terminating DTE
- Permanent virtual circuit (PVC)—A permanent path that never changes unless redesigned
- Switched virtual circuit (SVC)—A temporary path through the network established for the duration of the call and then dropped

X.25 can consist of PVCs and SVCs. A PVC could be used as a transmission path between two buildings. SVCs are generally used for switched-circuit transmission configurations. An X.25 network can be a combination of PVCs and SVCs (Figure 6.2).

X.25's delivery confirmation procedure specifies that the DTEs are to control the maximum number of data packets transmitted in accordance with network performance. A delivery confirmation bit set to 1 returns an acknowledgment to inform the sender that the data was received correctly.

If an error is detected by a terminating DTE, a retransmission is requested from the originating DTE. If the X.25 network detects errors, both the originating and terminating DTEs are informed and the data is re-sent.

X.25 Pricing

X.25 pricing is not distance sensitive (similar to dedicated line pricing) but packet sensitive. The more packets sent, the higher the cost.

FRAME RELAY SERVICES

Frame Relay service is another virtual private networking service. Frame Relay was developed to provide a high bandwidth wide area network without the need for long-term dedicated services. Packet-switching technology, similar to X.25, is again used. Frame Relay services take advantage of low-error signal transmissions by means of fiber transmission links and high-performance signal processing.

Less hardware is needed for each location than for a dedicated line, making installation relatively simple.

A Frame Relay network consists of all PVCs. They are monitored by a telecommunications carrier 24 hours a day, 7 days a week. Because a Frame Relay network is shared among many customers, there is a need to rely on the telecommunications carrier's policy not to oversell Frame Relay network capacity. If a Frame Relay network is oversold, frames may be dropped. If a customer loses enough data, he or she may consider leaving the service.

Frame Relay transfer speeds start from 56 Kbps and continue up to 45 Mbps. A Frame Relay access device (FRAD) is needed to connect to a Frame Relay service. The FRAD unit acts as an interface between the DTE and the Frame Relay network and is usually installed as a card within a router (Figure 6.3).

TIMETABLE OF HISTORIC FRAME RELAY SERVICE EVENTS

1991	First public Frame Relay network becomes operational.
1993	First satellite transmission for Frame Relay services is made.
1994	Frame Relay customer reporting and traffic analysis are provided.
2000	45 Mbps or Fast Frame Relay services are available.

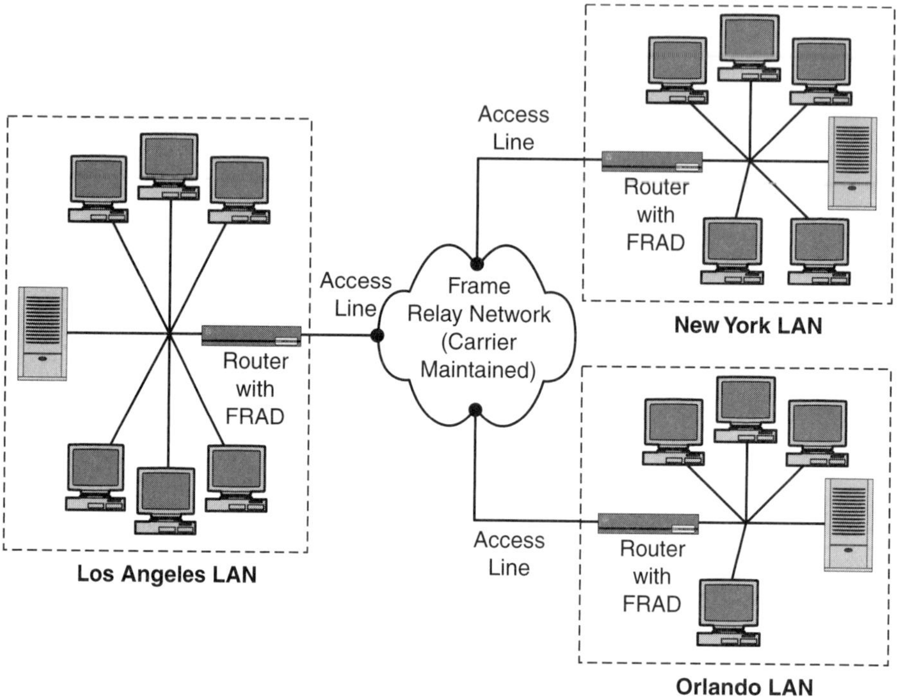

Figure 6.3
Frame Relay network.

Frame Relay Architecture

One of the main selling points regarding Frame Relay service is its ability to provide faster speeds if the network allows it. This ability to automatically increase the speed (depending upon network conditions) is called the burst rate or bursting. The standard committed rate and burst rate are all part of congestion control. Congestion control in Frame Relay networks includes the following elements:

- Admission control—Decision by the network whether to accept a new connection request, based on the relation of the requested traffic descriptor and the network's residual capacity.
- Committed information rate (CIR)—The average rate (in bps) at which the network guarantees to transfer payload data.

- Burst rate (bursting)—The maximum speed at which the frames can be transmitted during an interval.

Each telecommunications carrier must manage its own Frame Relay network. Virtual circuits provide access for a frame to route to its final destination. Due to the phenomenal growth of the Internet, businesses now demand increased speed and flexibility from their Frame Relay provider.

Some examples of Frame Relay service are restaurants, gas stations, and supermarkets. Franchised local establishments can connect with each other using Frame Relay service, exchanging critical data with a main office enabling management of multiple locations.

Frame Relay Pricing

Frame Relay pricing is neither distance sensitive nor packet sensitive. Instead, Frame Relay pricing contains several elements as follows:

- Committed information rate (CIR)—The guaranteed Frame Relay network speed
- Burst rate—The highest network speed rate (not guaranteed)
- Port speed rate—The speed of the access port providing entry into the network (node)
- Access circuit rate—The speed of the local access circuit provided by the LEC

Frame Relay pricing for CIR, burst rate, and port speed rate are all based on bandwidth only. The access circuit rate is based on both distance and bandwidth. For example, a New York customer orders an access line between the office and a telecommunication carrier's Frame Relay node at a T-1 rate. The port speed or node rate has been requested for 384 Kbps. The CIR has also been requested for 128 Kbps.

It is possible to burst up to the full 384 Kbps rate when the Frame Relay network can accommodate the additional speed; otherwise, the CIR of 128 Kbps will be used. In this example, the limiting factor is the port speed. If the New York customer wanted to increase the port speed to 768 Kbps, the burst rate would increase also (Figure 6.4).

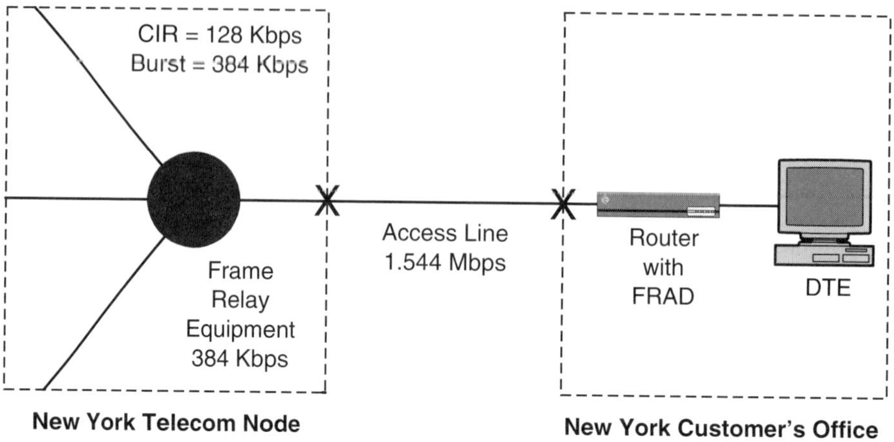

Figure 6.4
Frame Relay architecture and pricing.

Frame Relay Datagram Structure

Now that we understand some basic concepts of Frame Relay, let's analyze one frame individually. Each Frame Relay datagram consists of the following fields:

- Flag—The flag field synchronizes high-level data. This field designates the beginning and end of the frame with a unique bit pattern.
- Header—The header consists of the data link connection identifier bits (DLCI). The DLCI identifies the virtual connection. The receiving end knows which information connection this frame belongs to. Therefore, DLCI has only local significance.
- User data—This is the transmission or payload data. The maximum length of this data can be negotiated at frame setup time.
- Frame check sequence (FCS)—The FCS field implements error detection at each switching node. Error detection avoids bandwidth waste from the transmission of frames with errors.

The Frame Relay frame format begins with a starting flag, a header, user data, a frame check sequence field, and an ending flag (Figure 6.5).

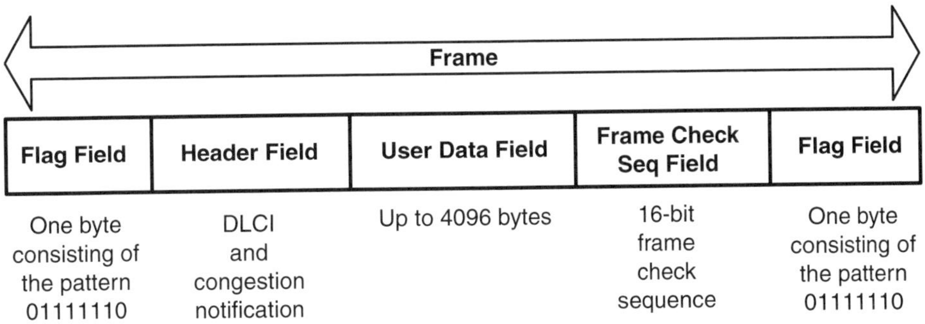

Figure 6.5
Frame Relay frame format.

FRAME RELAY AND X.25 COMPARISONS

X.25 provides error-free delivery by using error recovery techniques. Frame Relay takes advantage of today's fiber transmission networks, enabling it to obviate many of the services provided by X.25. The elimination of X.25 functions, combined with fiber transmission, enables Frame Relay to operate at speeds over ten times greater than that of X.25.

X.25 is designed to transmit packets, whereas Frame Relay is designed to transmit frames. X.25 packets contain bits dedicated for error and flow control. No error and flow control is needed for Frame Relay. The frames in Frame Relay contain an expanded payload data section that enables Frame Relay nodes to direct frames to their final destination. This occurs with a minimum amount of processing. X.25 has a fixed bandwidth available for transmission. It is a static form of transmission that uses or wastes bandwidth as necessary.

Frame Relay, on the other hand, is a dynamic form of transmission that can reallocate bandwidth during frame setup negotiation. Frame Relay networks transmit synchronous frames of data routed to a destination on the Frame Relay network.

Frame Relay is similar in principle to X.25 transmission. X.25 guarantees data integrity at the cost of network delays. Frame Relay switches packets end-to-end much faster, but there is no guarantee of data integrity at all.

Frame Relay is cost effective, partly because the network efficiencies are carefully optimized. Frame Relay is also much faster than X.25. The frames are sent to their destination with only a few milliseconds delay, as opposed to several hundred milliseconds delay with X.25.

VPNS FOR SMALL BUSINESS

VPN service is a private IP network maintained by a telecommunications carrier. There are different ways of implementing VPN services.

First, a VPN can be among a shared community (different companies). By sharing a VPN, companies can minimize costs. Users can connect across the network only to stations defined within their user group.

Second, a VPN can be available for one company without sharing. This provides maximum security with some additional cost.

Third, a VPN can use the public Internet for the transmission of packets. Some form of tunneling protocols will be necessary for security.

If a VPN seems like overkill for your small business, then Frame Relay services may be the answer. Frame Relay maximizes speed and throughput with significant cost savings. A Frame Relay network can provide wide area network connectivity with low setup costs.

CONCLUSIONS

This chapter covers another aspect, called virtual networks, of telecommunications services. The rise of data networks and the Internet has spawned VPNs and Frame Relay networks for both small and large businesses.

We learned that X.25 protocol was once a dominant player for packetized services. X.25 is now used for specialized financial transaction services.

We also discussed some aspects of security. Security should always be a consideration when these networks are reviewed. In many cases, the networks are shared. IP or Frame Relay data is transmitted and so can be captured by hackers. It is also possible that legitimate virtual network users can attempt to capture unauthorized files, leaving a company open to scrutiny or vulnerability. It is best to use some caution when analyzing virtual networks for your business.

Part 2

Telecommunications Voice and Data Terminal Equipment

7 Personal Computing Hardware

The personal computer (PC) is such an integral part of our lives today it is hard to understand how we survived without it. Can you imagine that in the not-too-distant past most small businesses had to create their financial balance sheet by hand (horrors!) or you would type your letter on an IBM Selectric typewriter (more horrors!). Oops, I forgot the Wite-out! I can go on and on, but it's clear to all of us that the personal computer is here to stay. We can't go back. Like it or not, we must deal with the PC and its benefits and pitfalls.

Today, the PC functions as an entertainment system, a business system, and a communications system. In this chapter, we look at all aspects of personal computing, including peripherals and external devices needed for today's business.

We concentrate on the hardware aspects of personal computing. PC hardware incorporates both the internal and external devices needed to complete a personal computing system in a small business environment.

We start by examining the central processing unit (CPU) and memory devices. Motherboards, power supplies, modems, and keyboard/mouse components are discussed as well as optical CD-ROM and mass storage devices. Video output devices, sound cards, printers, and other external devices are also discussed. The second section deals with both internal and external interfaces, serial, parallel, USB, FireWire, and MIDI. The third section takes a look at external output devices such as printers and video monitors. The fourth and final section covers PC networking, including local area network (LAN) and client/server technologies.

Two types of PCs are on the market today. The generic PC or nonproprietary PC allows some flexibility with mixing and matching internal components to create a personal computing system. The proprietary PC or nongeneric PC requires internal parts and components that are designed for your particular personal computing system only. For example, if you purchase a Dell product, then you must buy Dell memory.

We'll start off with PC fundamentals and then look at PC peripherals and external devices.

Let's boot up now!

PERSONAL COMPUTER ARCHITECTURE.................

The personal computer (PC) has several basic components that are a requirement of any system. The central processing unit (CPU) is the "brains" of the system. Input data is fed to the CPU or microprocessor. The CPU handles the data by either moving the data or performing calculations on it. CPU memory stores instructions and data temporarily and sends the final result to an output device. The memory bus provides the transmissions of signals between the CPU and memory, and the input/output bus, or I/O bus, provides both input and output transmission to and from the CPU. The motherboard provides the CPU support circuitry to process this data (Figure 7.1).

A serial port can connect I/O devices, such as a modem, a mouse, or a keyboard, directly to the CPU. Another I/O device such as a hard drive can store programs and data for future retrieval.

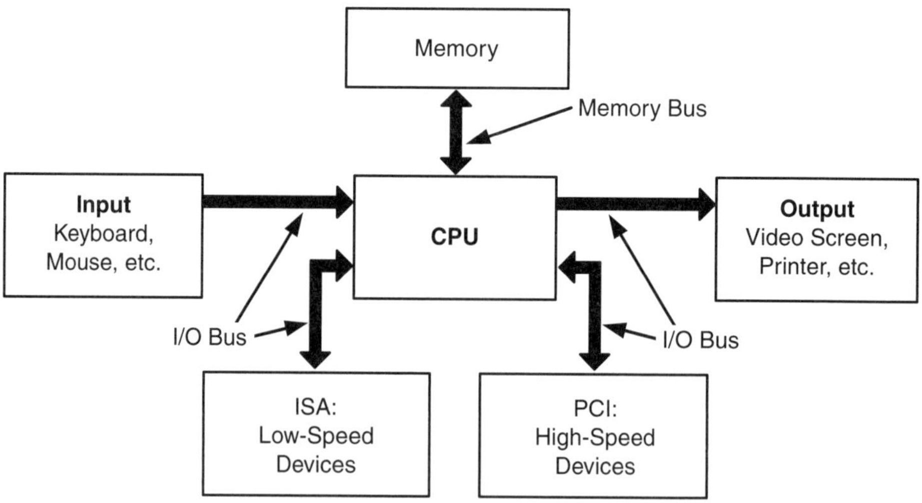

Figure 7.1
CPU interface.

Timetable of Historic Personal Computer (Hardware) Events

3000 BC	Abacus provides computations with sliding beads on a rack. This device is used for basic trading transactions.
1642	Blaise Pascal invents a numerical wheel calculator to help his father with tax collection duties.
1822	Charles Babbage invents a steam difference engine that performs calculations and prints results.
1889	Herman Hollerith develops the concept of punch cards to store data. After several mergers, his company became known as International Business Machines (IBM) in 1924.
1931	Vannevar Bush develops a mechanical calculator for solving differential equations.
1944	Howard Aiken produces a calculator, called the Mark I, that uses electromagnetic relays to signal mechanical parts.
1944	The Electronic Numerical Integrator and Computer (ENIAC) is developed. It consists of 18,000 vacuum tubes and consumes high levels of power and much physical space.
1945	The Electronic Discrete Variable Automatic Computer (EDVAC) is developed. It uses a stop/start or stored memory technique.
1951	The Universal Automatic Computer (UNIVAC) uses a central processing unit and predicts the winner of the 1952 presidential election.
1964	The International Business Machine (IBM) model 1401 mainframe computer processes financial information, using transistor technology.
1975	Popular Electronics magazine provides kit plans for the Altair computer utilizing the 8080 microprocessor.
1975	Paul Allen and William Gates develop Basic programming language for the Altair. Their new company is called Micro-Soft, later changed to Microsoft.
1976	Steven Wozniak and Steven Jobs develop a computer circuit board called Apple I.
1977	Commodore Business Machines markets the PET computer.
1977	Radio Shack Corporation markets the TRS-80 microcomputer with a Z-80 microprocessor, 4 KB of random access memory (RAM), and 4 KB of read-only memory (ROM).
1978	Texas Instruments (TI) introduces the TI-99 personal computer, which uses a 16-bit microprocessor.

Timetable of Historic Personal Computer (Hardware) Events (Continued)

1980	Hewlett Packard (HP) introduces the HP-85 microcomputer featuring 16-KB RAM.
1980	Sinclair Research markets the ZX-80 minicomputer, which uses a 780 eight-bit microprocessor.
1980	Seagate Technologies creates the first 5.25-inch hard drive storing 5 MB of data.
1981	Microsoft begins to develop a graphical user interface (GUI). This is the beginning of Windows operating system (OS) software development.
1981	IBM presents the 5150 Personal Computer (PC) in New York featuring a 4.77 MHz Intel 8088 microprocessor with 64-KB RAM.
1982	Commodore Business Machines markets the Commodore 64 microcomputer featuring a 6510 processor with 64-KB RAM.
1982	Atari introduces the 1200XL home computer with 64-KB RAM.
1983	IBM announces the next generation PC, called the XT, featuring a 10-MB hard drive and 128-KB RAM.
1983	Microsoft introduces the first mouse pointing device.
1983	Iomega introduces a removable storage device called the Bernoulli box.
1983	IBM announces PCjr featuring an 8080 microprocessor and 64-KB RAM.
1984	IBM introduces a portable PC with an Intel 8088 processor and 256-KB RAM.
1984	Hewlett Packard markets the first laser jet printer featuring an 8-page-per-minute (PPM) speed.
1986	Compaq Corporation introduces the first 16 MHz Intel 30386-based PC.
1987	IBM introduces the PS/2 line, which is their first 386 family microprocessor PC.
1987	IBM introduces video graphics array (VGA), providing 256 or 16 simultaneous colors.
1988	NEC Technologies introduces a 4.2 pound ultralight laptop PC.
1991	Intel Corporation introduces the 20 MHz 486sx family microprocessor. Note that the 486dx contains a math coprocessor.
1992	IBM introduces the Think Pad 700C laptop computer with a 25 MHz 486sl microprocessor.
1994	Commodore Business Machines files for voluntary liquidation of its assets.

The Central Processing Unit

The central processing unit (CPU) is the brains behind the personal computer (PC). The CPU directs all of the activities needed for data processing. Each CPU contains an Arithmetic Logic Unit (ALU). The ALU executes the instructions for the CPU. Central processing registers store the data temporarily until they can be processed by the ALU. The larger the register, the more efficient the overall CPU is. Some registers such as the Floating-Point Processing Unit (FPU) contain larger special-purpose registers to handle more complex number crunching.

A memory cache determines which memory address locations are used to store data for processing. Some of the newer Pentium microprocessors contain both level-1 and level-2 cache needed for faster clocking speeds.

Intel has established their own CPU family tree (Table 7.1). The x in the CPU name (80x86) designates the microprocessor generation. 80386 is the third generation, 803486 the fourth generation, and so forth. Intel decided not to create an 80586 fifth generation, instead calling the new series Pentium for trademark protection.

Table 7.1 Intel CPU Clock Speed

Generation	Processor	Speed	Comments
1st	8068, 8088	4 MHz	Used in first IBM PC
2nd	80286	8 MHz	
3rd	80386, 80386dx	33 MHz	
4th	486dx, 806dx2,4	100 MHz	
5th	Pentium	300 MHz	
6th (part#1)	Pentium Pro,P2	550 MHz	Integrated L2 cache
6th (part#2)	Pentium III Pentium IV	1 GHz 1.4 GHz	Full-speed L2 cache Net-Burst
7th	Itanium	800 MHz	

Each processor generation has a different form factor or mounting socket making interchanging CPU for each generation impossible (Table 7.2).

Table 7.2 Intel CPU Form Factors

Generation	Processor	Form Factor
1st	8068, 8088	40-pin DIP socket
2nd	80286	68-pin PGA socket
3rd	80386, 80386dx	132-pin PGA socket
4th	486dx, 806dx2,4	168-pin PGA socket 1 238-pin PGA socket 2 237-pin PGA socket 3
5th	Pentium	273-pin PGA socket 4 296-pin PGA Pentium MMX 320-pin PGA socket 6 321-pin PGA socket 7
6th (part#1)	Pentium Pro,P2	387-pin PGA Pentium ProSlot 1
6th (part#2)	Pentium III	370-pin PGA, Slot 1, Slot 2

I don't want to ignore Advanced Micro Devices (AMD) CPUs. AMD began producing a product starting with Intel's 80486 generation microprocessor. Today, AMD microprocessors are certainly on a par with Intel microprocessors at reduced prices.

AMD uses a "Socket 7" form factor for their K5 (Pentium) series. Their K6 (Pentium II) uses a "Super 7" form factor, and the K7 or Athlon (Pentium III) uses a "Slot A" form factor similar to Intel's "Slot 1."

Both Intel and AMD have produced overdrive CPUs that were newer generation CPUs designed to fit into older generation form factors and motherboards. Although this may seem like a great way to preserve an existing motherboard and other components, it may not always be cost effective. Advancements in technology for motherboards, PC components, and CPUs occur at such a rapid pace today it may be more cost effective to replace the entire system all at once. For this reason, I don't recommend using overdrive CPUs.

The design of PC motherboards will generally allow you to upgrade a CPU as long as form factor sockets and the CPU generation are the same. If you are ever required to replace a CPU, be sure to use an electro-static discharge (ESD) wrist strap, antistatic bags, and an antistatic discharge mat. These tools prevent static or a spark when you are handling sensitive CPUs.

One more issue regarding the CPUs: They run hot! If the CPU is not adequately cooled, it can literally burn out. To prevent a catastrophe, you must be sure that a heat sink is attached to the CPU with bonding glue or clips. The heat sink is basically a large piece of metal that can dissipate heat. A fan designed for CPU cooling attached to a heat sink can dissipate heat even faster.

TIMETABLE OF HISTORIC CENTRAL PROCESSING UNIT EVENTS

1948	The transistor is invented at Bell Laboratories.
1958	Jack Kilby, an engineer with Texas Instruments (TI), invents the integrated circuit (IC) microprocessor, combining three components on a single disk.
1972	Intel Corporation produces an 8008 200 KHz microprocessor that accesses 16 KB of memory.
1974	Intel produces an 8080 2 MHz microprocessor that accesses 64 KB of memory.
1984	Motorola Corporation creates the 68020 microprocessor with a clock speed of 16 MHz.
1984	Intel distributes the 80386dx family of microprocessors, which uses a clock speed of 16 MHz.
1987	Motorola ships the 68030 microprocessor.
1988	Intel introduces the 386sx microprocessor, which uses a 16 MHz clock speed and a 16-bit bus.
1989	Intel produces a 486 family microprocessor operating at clock speeds of 25 MHz. A 387 math coprocessor is integrated into this microprocessor.
1990	Motorola produces the 68040 microprocessor that has a clock speed of 25 MHz.
1992	Intel introduces the 486-dx2 family of processors with clock speeds of 25 MHz and 50 MHz.
1993	Intel introduces details of Pentium processor technology that has clock speeds of 60 MHz and 66 MHz.
1993	IBM and Motorola announce the PowerPC platform, which uses 50 MHz and 66 MHz clock speeds.
1994	Intel ships the dx-4 family of microprocessors with 2x, 2.5x, and 3x the clocking rates of previous 486 family microprocessors.
1994	Digital Equipment Corporation introduces its Alpha AXP processor with clock speeds of 300 MHz.
1994	Dr. Thomas Nicely at Lynchburg College discovers that Intel's Pentium microprocessor sometimes produces flawed floating-point decimal results.

Timetable of Historic Central Processing Unit Events (Continued)

1996	Intel creates the 200 MHz Pentium processor.
1996	Cyrix Corporation distributes the 133 MHz GX processor.
1997	Advanced Micro Devices unveils the K6 microprocessor at 166 MHz, 200 MHz, and 233 MHz processing speeds.
1997	Intel produces the 333 MHz Pentium II microprocessor with a 66 MHz bus speed.
1998	Intel releases the 350 MHz and 400 MHz Pentium II processors with 100 MHz bus speed.
1999	Intel distributes the Pentium III processor with a unique identity code.
1999	Advanced Micro Devices announces the Athlon microprocessor.
2000	Intel and Advanced Micro Devices develop microprocessors running at 1.1 GHz clock speeds.

Personal Computer Motherboards and System BIOS

The PC motherboard is by far the critical component of the entire system. You can upgrade a CPU, but if the memory and system bus are too slow, you are wasting your money. Each motherboard contains a chipset that acts as an interrupt controller for devices to access the CPU. Motherboards also contain a battery that can maintain memory settings for up to 12 years.

PC motherboards also contain the CPU, Random Access Memory (RAM), and interfaces to the various internal and external components. Processors may be faster, memory may be larger, but the essential components of a motherboard remain the same. When considering a major PC modification, you should determine how old your existing motherboard is, and what the CPU, memory, and bus speeds are before making a financial decision to replace the entire system or just the basic components.

The first motherboard standard was the IBM-PC AT standard. The AT motherboard was 12 inches wide and had only one keyboard input port connected directly on the board. All access for other input and output ports was through connectors that required cables to join with other connectors at the back of the case. The original AT motherboard bus speed was 4 MHz, which was soon upgraded to 8 MHz.

The next standard motherboard form factor was the ATX motherboard, which was developed by Intel in 1995. This motherboard was much smaller than its AT predecessor, and the positioning in the case was also different. For the first time, cooling fans were reversed, allowing air to flow inward instead of outward. Input/output connectors were now part of the motherboard, and a specially designed ATX case was needed to accommodate the change in design.

Newer form factors, such as LPX and NLX motherboards, are emerging. The goal of their redesign is to provide the same functionality as an ATX motherboard in a smaller package.

Another important aspect of the motherboard is the system read-only memory (ROM) BIOS. The BIOS is capable of storing small programs that the CPU constantly uses to interpret components and devices within the system itself.

Upon system boot-up, the system BIOS is the first set of instructions the PC uses.

Today's BIOS can be upgraded from the manufacturer by means of erasable programmable read-only memory (EPROM) circuits and Flash EPROM. EPROM circuitry allows the set of system BIOS instructions to be reprogrammed.

EPROM is generally slow, probably too slow, which is why many manufacturers use Flash EPROM today. Manufacturers can post upgrades to system BIOS on their web sites for users to upgrade Flash EPROM. A word of caution, however, is that any corrupted or garbled data entering the Flash ROM (noise on the line, crosstalk, etc.) can render the EPROM useless. Once the EPROM is not working or corrupted, it cannot be reprogrammed and must be permanently removed from the motherboard.

CPU Memory

CPU memory can be as important as the CPU itself. As CPUs become faster, memory must become faster to keep up with the pace. If system memory is too slow, performance will suffer and the CPU will work harder even with the latest and greatest CPU. During the PC boot-up process, memory must be detected for the machine to operate successfully.

Generally, when we speak of computer memory we are speaking of random access memory (RAM). Although there are other types of memory, RAM is the most important. You can never have enough RAM; 128 MB seems to be today's standard. If the CPU cannot find enough RAM available, it must use virtual memory (VM). VM requires space on your hard drive and is much slower than RAM.

Both programs and the data itself can be stored in RAM. It is important to note that RAM is volatile. It starts off with a clean slate as the PC starts and then cleans itself when the PC is shut down.

RAM may look different in its physical appearance, however; each memory address is part of a grid with rows and columns. To access data stored in memory, you must access the correct address of the grid. The refresh rates of these microchips are extremely fast. Some microchips have a 4K or a 2K refresh rate. A 4K refresh would have the equivalent of 4,000 address columns. The capacity of a microchip is the rows multiplied by the columns, which are then multiplied by the bit width. Some microchip manufacturers refer to the capacity as the density.

Microchip manufacturers have changed the physical appearance of memory over the years, first starting with basic dual inline pin package (DIPP) sockets and then progressing to single inline memory module (SIMM) packages and dual inline memory module (DIMM) packages. Physically, the SIMM package has a single notch on its lower edge and the DIMM package has a dual notch on its lower edge, making them easily distinguishable.

Memory modules (SIMM or DIMM) within a memory bank (group of SIMM or DIMM) must be identical. In other words, you cannot mix and match. It is also a good practice to keep memory speeds identical. Memory slots get filled starting from bank 0 (first slot) on up with no spaces in between (Table 7.3).

Table 7.3 Memory

Module	No. of Pins	Bit Width
DIPP	—	1-bit
SIMM	30	8-bit
SIMM	72	32-bit
DIMM	168	64-bit

All types of memory, including RAM, are subject to errors. While substantial progress has been made in keeping memory errors to a minimum, errors do exist. In many cases, a single-bit error is not very important especially for pixels on a video screen or printer. However, an error in a calculation could prove to be catastrophic and cause the entire CPU to malfunction and require boot-up.

For this reason, the concept of parity or error checking was developed. Parity is simply an extra bit added to an 8-bit byte. By counting the number of 1's in an 8-bit byte, you will obtain a result that is either odd or even. For example, a 11011100 byte will obtain a result of 5, which is an odd number. Therefore, the parity bit will be set at 0. On the receiving side, the 1's are counted again and the parity bit is checked. If the parity bit doesn't match, there is an error in the data.

There are two types of parity: odd parity and even parity. You cannot mix parity within a memory bank. If you are adding more memory, you must use the same parity as that of the SIMM or DIMM already installed.

The most obvious problem with parity checking is the possibility that there will be an undetected double error (2-bit error) with a matching parity bit. This will be viewed as if no error exists. The other problem occurs when parity error is detected. In this situation, the CPU is instructed to shut down so data will not be further corrupted.

An improvement over parity error checking is error correction code (ECC). ECC can detect memory errors and then attempt to fix them by using complex algorithms.

Hard Drives and Other Magnetic Storage Devices

Random access memory holds programs and data for a relatively short period of time. RAM can be considered a temporary workspace that is designed for quick execution of instructions and data output. RAM is sometimes referred to as volatile storage. That is, RAM doesn't retain its data once the power is turned off.

A magnetic storage device is designed for a different purpose. This device, frequently known as a hard drive, retains large amounts of data according to the principles of electromagnetism. The storage media contains a movable surface that can store a magnetic charge. This storage media can be rigid or flexible. A hard drive usually has an aluminum or glass surface, which spins, while another magnetic storage device, called a floppy drive, contains a flexible Mylar surface, which also spins.

A tape recorder with cassette tape or reel-to-reel tape works on the same principles as a hard drive; however, the media is usually designed for analog audio.

It is possible to create your own magnetic storage device by taking a strip of adhesive tape and gathering some iron oxide (rust). Yes—rust. You can easily obtain rust from metal outdoor furniture or metal backyard fencing. Scrape the rust off the metal fence or furniture into a small pile, keeping the particles as small as possible. Sprinkle the iron oxide onto the adhesive tape and then allow the tape to pass over a magnetic head. A reel-to-reel tape recorder works best. Now record your voice into the tape recorder through a microphone while the adhesive tape containing iron oxide passes over the magnetic head. When you rewind the tape and play back the sound, you will hear your voice, although it will sound rather crude, I'm sure. The reason is that the magnetic tape is manufactured under tight quality control specifications which cannot be duplicated at home.

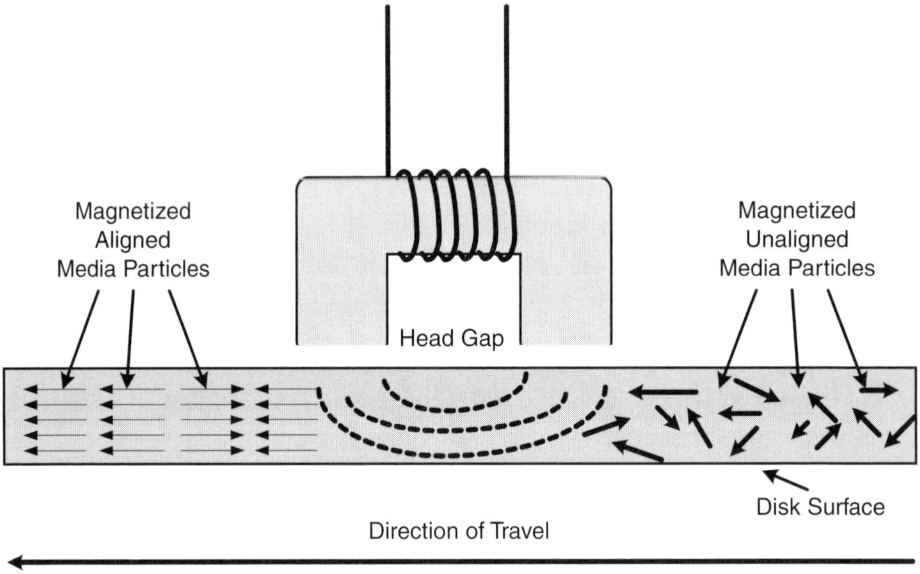

Figure 7.2
Writing to a hard drive.

We now understand the materials needed to create an electromagnetic effect. However, we still need to understand what it takes to store digital information on a floppy disk or hard drive.

An unformatted hard drive contains randomly stored particles that are disorganized. There is no data to read because the magnetic particles are not aligned in any fashion. For data recording and particle alignment, a write head, which remains very close to the rotating media surface (Figure 7.2), is used.

An electrical charge applied to the read/write head will align all of the magnetized media particles in the same direction. If the charge on the head is positive, the particles will be aligned in one direction. If the charge on the head is negative, the particles will be aligned in the opposite direction.

As the disk surface moves and the head passes over sections of the media, a pattern of aligned particles is created. The direction or charge of the magnetized particles determines the bit storage of a 1 or a 0.

Reading information off a disk is similar to writing to a disk but in the opposite sense. The electromagnetic charge from the movable media is picked up by the read/write head, which generates a small current and voltage based on the direction or charge of the aligned particles. The direction or charge of this small current determines a 1 or a 0.

Hard drive manufacturers continually strive to squeeze more and more charged particles into a smaller and smaller space. To achieve the goal of a physically small hard drive with gigabits of data storage, IBM developed magneto resistive (MR) technology. This technology requires a separate MR head to constantly monitor the current to ensure error-free operation.

Another important aspect of hard drive operation is timing. Without proper timing, the head will misread the aligned particles and produce errors.

The hard drive device is not something you should attempt to open and repair. Hard drives are manufactured in environmentally friendly, dust-free rooms to keep them virtually error free.

Physically, all hard drives contain several platters. Each platter is the spinning rigid media made up of aluminum or glass that we discussed earlier. The spinning motion of the platters and the in-and-out motion of the heads provide quick access to an address or track on each of the platters (Figure 7.3).

Today's hard drives use a voice call actuator that provides precise adjustment of the actuator arm. The voice call actuator moves the arm in and out using electromagnetic attraction and repulsion. There is also a feedback loop circuit to fine-tune the actuator arm if necessary.

The spindle motor is directly attached to the spindle shaft. This is important because there are no belts or pulleys to wear out. This motor rotates at high speeds: usually at 7,200 revolutions per minute (rpm) or 10,000 rpm. The higher the speed, the less time it takes to retrieve information.

Each hard drive has a built-in logic board. The logic board is itself a sophisticated computer with its own CPU and memory. The hard drive logic board performs the following functions:

- Monitors and adjusts spindle motor speed
- Monitors and controls timing
- Amplifies read signals from the read/write heads to acceptable levels
- Provides a feedback loop circuit to monitor the position of the actuator arm and adjusts if necessary

Because of the mechanical nature of hard drives, these devices are susceptible to failure. Care should be taken to avoid any extreme conditions such as excessive temperature or g-force (dropping, bumping, etc.). I cannot stress enough the importance of backing up important data. Without a backup plan in place, it is possible to lose all data; that loss can have a detrimental effect on your PC or small business in general.

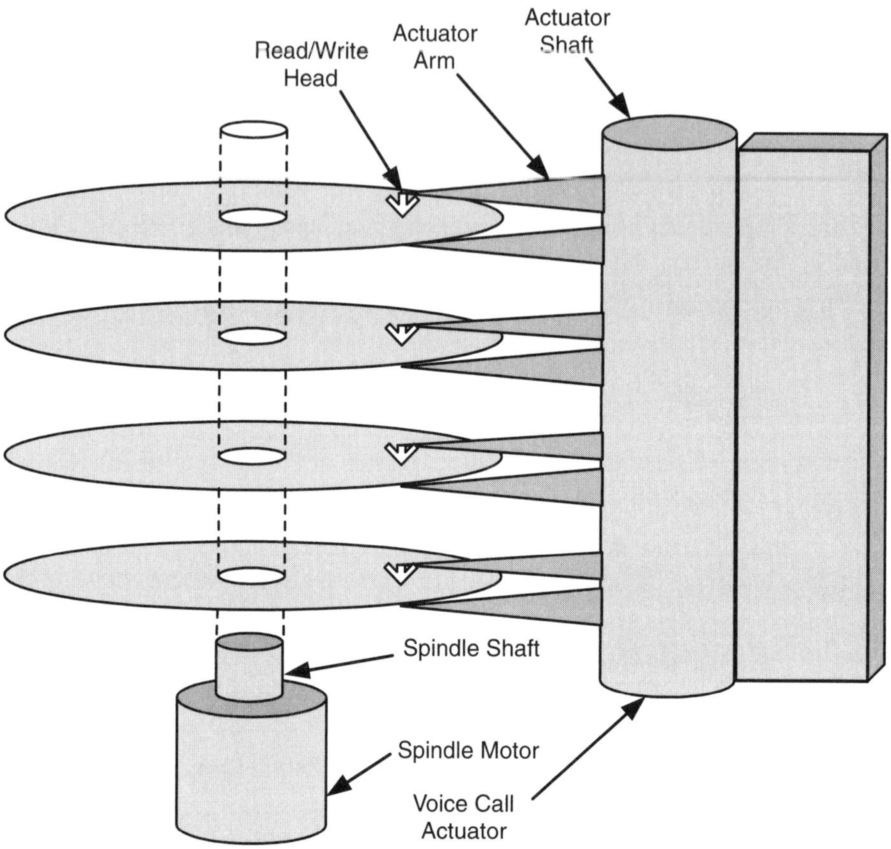

Figure 7.3
Hard drive operation.

Floppy Disk Drive

The floppy disk drive is similar to a hard drive with the following exceptions:

- Has limited storage space of 1.4 MB
- Uses a stepper motor instead of a voice coil actuator
- Uses a single soft Mylar media disk instead of hard media disks

Because of the size of programs used to load software, CD-ROMs (650 MB) have replaced floppy disks as a software program storage device. However, floppy disk drives are still included with PCs for the following purposes:

- Alternate system boot-up disk
- Small, inexpensive, removable data file storage
- Distribution of software drivers needed for equipment installation

If you are looking for a removable drive that can store larger amounts of data, you may find a cartridge drive to be useful.

Cartridge Drive

A cartridge drive such as the Iomega Zip can store 100 MB of data in a cartridge about twice as thick as a floppy drive. Again, the principle is the same as a floppy drive. In fact, many manufacturers now include internal Iomega Zip drives with their PCs as a standard package.

Unfortunately, even 100 MB of removable storage is not enough for today's software programs and high-end graphics, so the Iomega Jaz was created. Iomega Jaz cartridges come in 1-GB and 2-GB sizes that are interchangeable.

A 1-GB or 2-GB cartridge is great for data backup. In fact, Iomega includes automated data backup software with its product to make this task even easier. It is a good idea to create a long-term backup cartridge that can be stored off-site each month and a short-term backup cartridge that can be created each week.

One other fact about cartridges is that they can be expensive. A 2-GB cartridge can cost ~$100, so plan your removable storage needs accordingly.

We've discussed the cartridge drive, floppy drive, and hard drive (Figure 7.4). Now let's take a look at optical storage devices, which can provide long-term data storage also.

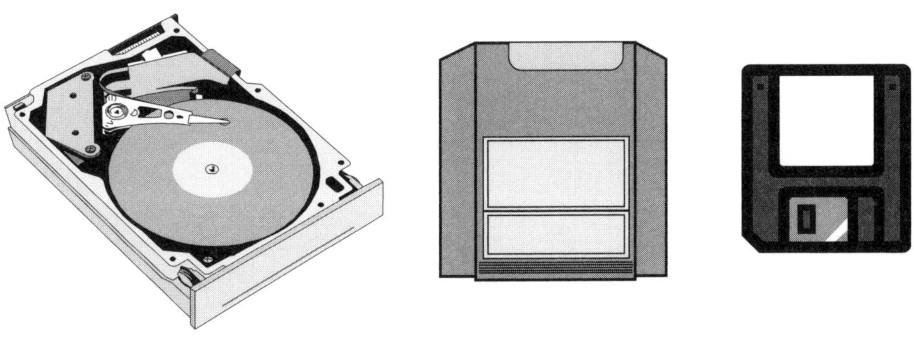

Figure 7.4
Hard drive, cartridge drive, and floppy drive.

CD-ROM Optical Storage Devices

We've already mentioned how the floppy disk has been replaced by the CD-ROM for software installation and large data storage. CD-ROMs are optical storage devices that use a laser beam instead of a magnetic pickup head to read and write data. There are some similarities and some differences between hard drives and CD-ROM drives.

The spindle motor spins the CD-ROM disk while the laser head moves in and out reading the data. The laser shines a light on the reflective surface of the CD-ROM and a photodetector reads the reflected light. The disk itself contains lands and pits representing 1's and 0's. The lands reflect nearly all of the laser light back to the photodetector, whereas the pits dissipate much of the light and reflect less.

One of the benefits of optical storage is that there is no physical wear on the media. Only the laser light beam touches the media. I would also like to mention here that a CD-ROM laser can be somewhat dangerous if you shine it directly into someone's eyes. Generally, this doesn't happen because the CD-ROM drive is enclosed and is usually too expensive to repair. Generally, CD-ROM drives are not repaired because of their relatively low cost. Replacements of nonfunctioning CD-ROM drives are the norm.

The track density of a CD-ROM is generally higher than that of a hard drive, so a microcontroller voice call actuator is necessary. The seek time is slower than that of a hard drive. Hard drives are designed for fast retrieval of data, whereas CD-ROM drives are not quite as fast.

CD-ROM drives also work on the concept of constant linear velocity. The spindle motor speed on a 1X drive can adjust from approximately 200 to 540 rpm. The speed depends on where the data resides. If the laser head is positioned near the inside of the disk, the spindle motor speed will be slower. Conversely, if the laser head is positioned near the outside of the disk, the spindle motor speed will be faster. The reason is that CD-ROM drives read their data at a constant rate of 150 Kb/sec, starting from the inner track and moving outward.

Over a period of time, consumers demanded a faster read rate, so 4X, 8X, 12X, and higher speeds were developed. In fact, the spindle motors of a 48X CD-ROM run at speeds of 12,000 rpm to achieve a 7.2 Mb/sec transfer rate.

Compact Disk Formats

The CD-ROM format started with the audio compact disk (CD) that eventually replaced analog records back in the 1980s. CDs are digitized by a 44.1 kHz sampling rate at 16-bit samples in a stereo format. This standard, which is still used today, is the red book format containing 74 minutes of uncompressed digital audio (Table 7.4).

Table 7.4 CD-ROM Standards

Format	Description
Red Book	74 minutes of uncompressed digital audio
Yellow Book	CD-ROM format allowing for 650 MB of data storage
CD-ROM XA	Yellow book extension allowing for PhotoCD and VideoCD formats
Green Book	CD-1 format (no longer used)
Orange Book (Pt.1)	CD-R recordable multisession disk; magneto optical format
Orange Book (Pt.2)	CD-R recordable (write once read many)
Orange Book (Pt.3)	CD-RW rewriteable CD-ROM format
White Book	Video-CD format using MPEG-1 compression
Blue Book	Enhanced music CD containing mixed formats (audio and data)

CD-Recordable Drives

The CD-ROM drive is a playback-only medium. Until recently, CD-R drives either were not available or were outrageously expensive. In fact, the price is now so low that CD-R disks can be purchased for as little as $.50 each with the CD-R drive itself as low as $200. With these prices, the CD-R format is practicable for data backup and long-term storage of data in a Write Once Read Many (WORM) format.

The CD-R drive is similar to the CD-ROM drive with the exception of the laser. Once the data begins writing to the disk, you cannot stop the disk without ruining it. If the stream of data to the disk is interrupted for any reason, you can add your disk to the circular file (trash bin) and start over again with a brand new CD-R disk.

A CD-R disk itself is somewhat different from a standard CD-ROM playback disk. The CD-R contains a spiral pre-groove for the laser to follow. The pre-groove contains one giant land. The disk itself contains a layer of photosensitive dye, another layer of reflective material, and a protective plastic layer. The CD-R laser, more powerful than a standard CD-ROM laser, causes the dye layer to react with the metallic layer, creating a pit. So the description of "burning" a CD is true. You are literally causing a reaction between two layers and creating pits.

CD-R drives also have speeds of 2X, 4X, 8X, and 16X. The 16X CD-R version can copy a CD-ROM bit by bit in approximately 7 minutes!

CD-RW Drives

An improvement over the CD-R is the CD-RW, which can reuse disks by erasing old data and rewriting new data. The CD-RW disk itself contains an additional layer called the recording layer. The recording layer contains a special chemical which, when heated to 700° Celsius and then quickly cooled, creates pits. A lower-level heat at 500° Celsius will cause the recording layer compound to create lands, which are reflective.

A CD-RW disk must be read by a CD-RW drive. However, CD-RW drives can almost always create CD-R disks—a great feature. Because of the features associated with CD-RW drives, a CD-RW drive is sometimes called a multiwrite multiread drive.

Before we leave this topic, I don't want to forget about DVD-Recordable, or DVD-R format. Several different DVD-R formats exist today, all of which are very expensive. DVD-R is similar to a CD-R, which can hold 4.7 GB on a single layer disk. DVD-RAM operates similarly to CD-RW, except that a cartridge is used. The DVD-RAM disk is permanently sealed within the cartridge and is not compatible with any other DVD player.

CD-ROM vs. Digital Video Disk

Remember that in Chapter 4 we reviewed Digital Versatile Disk (DVD) technology and compared it with the Video Cassette Recorder (VCR). Well, this time let's com-

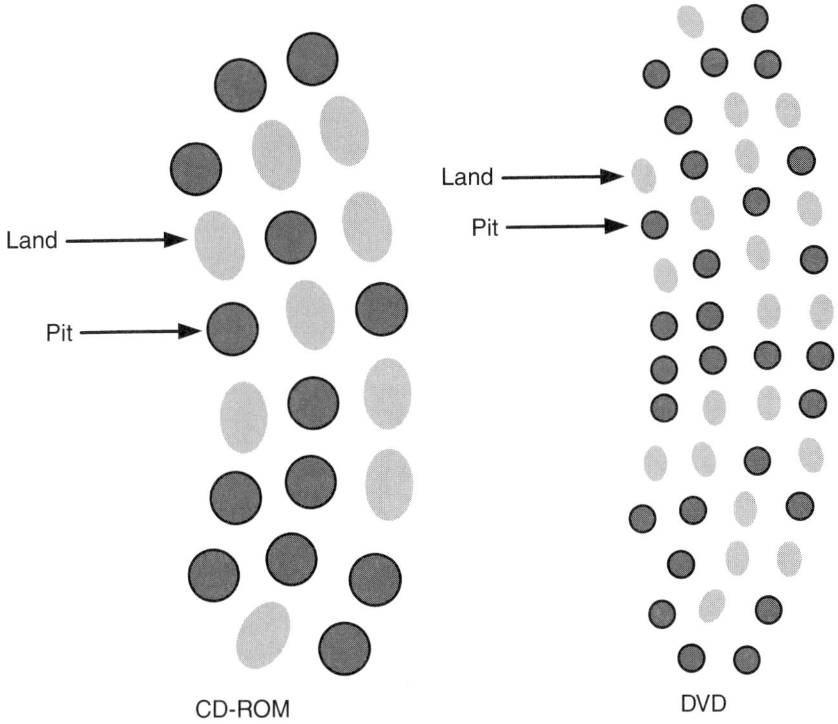

Figure 7.5
CD-ROM and DVD comparisons.

pare the DVD with the CD-ROM. A DVD uses smaller lands and pits than those on a CD-ROM disk, allowing for increased storage capacity (Figure 7.5).

DVD and CD-ROM are physically the same size, but they differ in several ways:

- Pits on a DVD (.4 micron) are approximately half the size of those on a CD-ROM (.834 micron).
- Tracks on a DVD (.74 microns) are much closer together than on a CD-ROM (1.6 microns).
- A DVD has an improved error checking code than that of a CD-ROM, providing more space for data on the disk.
- A double-sided DVD uses two complete disk assemblies attached back-to-back. You must physically turn over the disk to play the opposite side. CD-ROMs currently do not have double-sided disks.

- A DVD can also have two data layers. One layer would be opaque and the other layer would be translucent. The laser can focus on one layer or the other. CD-ROMs do not have this capability.

At the present time, most software is distributed on CD-ROM. However, there may be a time when manufacturers' software is distributed on DVD to provide increased space on a single disk.

Personal Computer Audio

Personal computer audio is a major component of today's integrated PC hardware. In fact, the audio section is frequently built on the motherboard itself. You will recall our discussion of Nyquist's sampling theorem and what it takes to convert analog audio into a digitized bit stream.

As a general rule, the higher number of bits used to store the sound digitally, the greater the accuracy in reproducing the sound (Table 7.5).

Table 7.5 Audio Quality

Quality Type	Bandwidth	File Size	Sample Frequency
Telephone	8-bit	11 KB/Sec or 660 KB/Min.	11,025 Hz
Radio	8-bit	22 KB/Sec or 1.3 MB/Min.	22,050 Hz
CD Quality (Stereo)	16-bit	172 KB/Sec or 10.3 MB/Min.	44,100 Hz

Pulse code modulation (PCM) is the standard format used to store audio data; the Windows file extension is .WAV. Generally, a 16-bit audio adapter with wavetable synthesis (see MIDI section) is adequate for most applications.

Creative Labs was one of the first audio adapter manufacturers for the PC with its Sound Blaster product. Today, it is the de facto standard for audio PC hardware.

With the availability of low-cost hard drives with sizes starting from 20 GB, it is possible to save audio notes, keep an audio diary, record a meeting, etc. Just be sure to keep your quality at the lowest acceptable levels to save on hard disk space.

CD audio music can be played through the CD-ROM with an audio adapter cable. The cable connects directly from the CD-ROM drive to the sound card, which amplifies the sound to the speakers. The cable contains four wires for stereo sound (two wires for each speaker).

Personal Computer Video

Personal computer video is one of the most important subsystems of the PC itself. The display of graphics and images frequently determines how well a computer performs. The PC video adapter card has its own memory and its own processor.

Screen resolution is one of the most important aspects of a PC video card and is expressed by the number of pixels displayed for height and width. The resolution you require is determined by the capabilities of your PC monitor. More about PC monitors in the next chapter.

So let's say you have a 21-inch monitor and use a screen resolution of 1024 x 768 pixels. Great! Now let's try the same resolution on a 14-inch monitor. The characters are very small on a 14-inch monitor with a 1024 x 768 pixel resolution. You may want to decrease the resolution on a 14-inch monitor to an 800 x 600 pixel resolution.

Another important aspect of PC video is the color depth. The color depth is sometimes described as the third dimension. The color depth provides varying degrees and mixtures of color for each pixel. For display of the color variations required for today's graphic intensive displays, multiple bits of memory are required for each pixel. This is why sufficient video memory is required to display higher resolutions (Table 7.6).

Table 7.6 Video Resolution and Memory Requirements

Video Resolution	Color Depth	Number of Colors	Memory Requirements
640 x 480 pixels	16-bit	65,536	614.4 Kbytes
640 x 480 pixels	24-bit	16,777,216	821.6 Kbytes
800 x 600 pixels	16-bit	65,536	960.0 Kbytes
800 x 600 pixels	24-bit	16,777,216	1.44 Mbytes
1,024 x 768 pixels	8-bit	256	768.4 Kbytes
1,024 x 768 pixels	16-bit	65,536	1.57 Mbytes
1,024 x 768 pixels	24-bit	16,777,216	2.35 Mbytes
1,280 x 1,024 pixels	4-bit	16	655.3 Kbytes
1,280 x 1,024 pixels	8-bit	256	1.31 Mbytes
1,280 x 1,024 pixels	16-bit	5,536	2.62 Mbytes
1,280 x 1,024 pixels	24-bit	16,777,216	3.93 Mbytes

Three video color depth standards have evolved over time:

- True Color—Highest quality with most memory used. 24 bits assigned to each pixel with red, blue, and green using 8 bits each.
- High Color—Moderate quality with a medium amount of memory used. 16 bits assigned to each pixel (5 bits for red, 5 bits for blue, and 6 bits for green).
- 256 Color—Lower quality with a lower amount of memory used. 8 bits assigned to each pixel sharing a 256-color palette.

Generally 8 MB of memory is sufficient for today's video display graphics. However, graphics accelerators and 3D video graphics adapter cards with 64 MB of video memory have been developed for the PC game industry.

One other aspect of video display is the refresh rate. The refresh rate is determined by the speed at which the video display card can operate to reduce or eliminate flicker. Flicker is easier to see on a larger monitor than on a smaller one.

In most cases, an interlaced video refresh is not used (recall our discussion of television displays in Chapter 4). An interlaced refresh mode refreshes half of the image at a time. The video display adapter and the monitor must work together, so an acceptable rate is usually determined by both devices. Usually (and thankfully!), refresh rates are determined automatically by the Windows OS and by the video display card manufacturer. Video refresh rates can be standardized as follows:

- 60 Hz or less—Produces a flicker that is noticeable
- 60 Hz to 72 Hz—Produces a somewhat noticeable flicker; however, 60 to 72 Hz rates are generally acceptable
- 72 Hz to 85 Hz—A generally accepted refresh rate
- 85 Hz and above—Be careful with this one. You gain nothing by setting the rates too high. In fact, attempting to drive a monitor at too high a rate can cause irreparable damage to the monitor, rendering it useless.

The output of a video display card is a 15-pin connector that connects to a video monitor (Table 7.7).

Table 7.7 VGA Pin Description

Pin	Description	Direction
1	Red Video	toward monitor
2	Green Video	toward monitor
3	Blue Video	toward monitor
4	Monitor ID 2	toward PC
5	Ground	—
6	Red Analog Ground	—
7	Green Analog Ground	—
8	Blue Analog Ground	—
9	Key	—
10	Sync Ground	—
11	Monitor ID 0	toward PC
12	Monitor ID 1	toward PC
13	Horizontal Sync Pulse	toward monitor
14	Vertical Sync Pulse	toward monitor
15	Monitor ID 3	toward PC

Video monitors, accelerated graphics port (AGP), and multimedia applications are discussed in a later section of this chapter. Now, let's take a look at input devices such as keyboards and mouse devices before we reexamine modems.

Personal Computer Input Devices

The keyboard and mouse have seen very little change since their inception with the first IBM XT computer in 1983. Today, there are enhanced keyboards with more characters and ergonomic keyboards that encourage a more natural hand position. Many keyboard manufacturers customize their keyboards and include additional features like volume control or browser functions like home or favorites.

A 6-pin miniconnector connects the keyboard to a PC (Table 7.8).

TIMETABLE OF HISTORIC KEYBOARD EVENTS

1868	Christopher Latham Sholes invents the "QWERTY" typewriter in Milwaukee. Letters are rearranged into positions where the keys won't jam.
1874	The first Remington #1 typewriter appears on the market.
1878	The Remington #2, which contains both upper and lower case, appears on the market.
1932	August Dvorak challenges the QWERTY keyboard by arranging his letters according to frequency. Dvorak's home row (second row) uses five vowels and five of the most common consonants "AOEUIDHTNS." The home row does 70% of the work on a Dvorak keyboard, whereas a standard typewriter's home row does 32%.
1953	A U.S. General Services Administration study finds that it doesn't matter what keyboard you use. Good typists can just type faster than bad typists.
1983	The IBM Personal Computer Model XT features a QWERTY keyboard.

In 1968, the American National Standards Institute (ANSI) coded an 8-bit standard still in use today. The American Standard Code for Information Interchange (ASCII) was created as a standard translation table between the alphabet and binary bits. The eighth bit is sometimes turned off or used for additional characters. Additional characters could be the non-English alphabet or symbols. If the eighth bit is used, the format is called extended ASCII (Table 7.9).

Table 7.8 Keyboard Pin Description

Pin Number	Comment
1	Keyboard data
2	(unused)
3	Ground
4	+5 V power
5	Keyboard clock
6	(unused)

Table 7.9 Binary, ASCII-Hexadecimal Conversion Chart

Binary	Hex	ASCII	Binary	Hex	ASCII	Binary	Hex	ASCII	Binary	Hex	ASCII
000 0000	00h	ctrl-@	010 0000	20h	(space)	100 0000	40h	(	110 0000	60h	`
000 0001	01h	ctrl-A	010 0001	21h	!	100 0001	41h	A	110 0001	61h	a
000 0010	02h	ctrl-B	010 0010	22h	"	100 0010	42h	B	110 0010	62h	b
000 0011	03h	ctrl-C	010 0011	23h	#	100 0011	43h	C	110 0011	63h	c
000 0100	04h	ctrl-D	010 0100	24h	$	010 0100	44h	D	110 0100	64h	d
000 0101	05h	ctrl-E	010 0101	25h	%	010 0101	45h	E	110 0101	65h	e
000 0110	06h	ctrl-F	010 0110	26h	&	010 0110	46h	F	110 0110	66h	f
000 0111	07h	ctrl-G	010 0111	27h	'	010 0111	47h	G	110 0111	67h	g
000 1000	08h	ctrl-H	010 1000	28h	(	010 1000	48h	H	110 1000	68h	h
000 1001	09h	ctrl-I	010 1001	29h	)	010 1001	49h	I	110 1001	69h	i
000 1010	0Ah	ctrl-J	010 1010	2Ah	*	010 1010	4Ah	J	110 1010	6Ah	j
000 1011	0Bh	ctrl-K	010 1011	2Bh	+	100 1011	4Bh	K	110 1011	6Bh	k
000 1100	0Ch	ctrl-L	010 1100	2Ch	'	100 1100	4Ch	L	110 1100	6Ch	l
000 1101	0Dh	ctrl-M	010 1100	2Dh	-	100 1101	4Dh	M	110 1101	6Dh	m
000 1110	0Eh	ctrl-N	010 1110	2Eh	.	100 1110	4Eh	N	110 1110	6Eh	n
000 1111	0Fh	ctrl-O	010 1111	2Fh	/	100 1111	4Fh	O	110 1111	6Fh	o
001 0000	10h	ctrl-P	011 0000	30h	0	101 0000	50h	P	111 0000	70h	p
001 0001	11h	ctrl-Q	011 0001	31h	1	101 0001	51h	Q	111 0001	71h	q
001 0010	12h	ctrl-R	011 0010	32h	2	101 0010	52h	R	111 0010	72h	r
001 0011	13h	ctrl-S	011 0011	33h	3	101 0011	53h	S	111 0011	73h	s

Table 7.9 Binary, ASCII-Hexadecimal Conversion Chart (Continued)

Binary	Hex	ASCII	Binary	Hex	ASCII	Binary	Hex	ASCII	Binary	Hex	ASCII
001 0100	14h	ctrl-T	011 0100	34h	4	101 0100	54h	T	111 0100	74h	t
001 0101	15h	ctrl-U	011 0101	35h	5	101 0101	55h	U	111 0101	75h	u
001 0110	16h	ctrl-V	011 0110	36h	6	101 0110	56h	V	111 0110	76h	v
001 0111	17h	ctrl-W	011 0111	37h	7	101 0111	57h	W	111 0111	77h	w
001 1000	18h	ctrl-X	011 1000	38h	8	101 1000	58h	X	111 1000	78h	x
001 1001	19h	ctrl-Y	011 1001	39h	9	101 1001	59h	Y	111 1001	79h	y
001 1010	1Ah	ctrl-Z	011 1010	3Ah	:	101 1010	5Ah	Z	111 1010	7Ah	z
001 1011	1Bh	ctrl-[	011 1011	3Bh	;	101 1011	5Bh	[	111 1011	7Bh	{
001 1100	1Ch	ctrl-\	011 1100	3Ch	<	101 1100	5Ch	\	111 1100	7Ch	|
001 1101	1Dh	ctrl-]	011 1101	3Dh	=	101 1101	5Dh	]	111 1101	7Dh	}
001 1110	1Eh	ctrl-^	011 1110	3Eh	>	101 1110	5Eh	^	111 1110	7Eh	`
001 1111	1Fh	ctrl-_	011 1111	3Fh	?	101 1111	5Fh	_	111 1111	7Fh	ctrl-del

Believe it or not, there was a time when a pointing device was considered unnecessary or exotic. The pointing device or mouse has become so ingrained into our computing culture that we cannot conceive of a computer without one. Interestingly enough, it took more than a decade before this simple but effective device was included on any type of computer.

Timetable of Historic Mouse Events

1957	Stanford Research Institute (SRI), which is funded by U.S. government grants to improve its superiority in technology, employs Douglas Engelbart.
1968	Engelbart and his group demonstrate an online system containing shared screen teleconferencing, hypertext linking, online help, word processing, and an online pointing device.
1968	Engelbart and his group call their pointing device a "mouse" because it looks like one by virtue of its "tail."
1984	A mouse is included with the IBM PC and the Apple Mac.

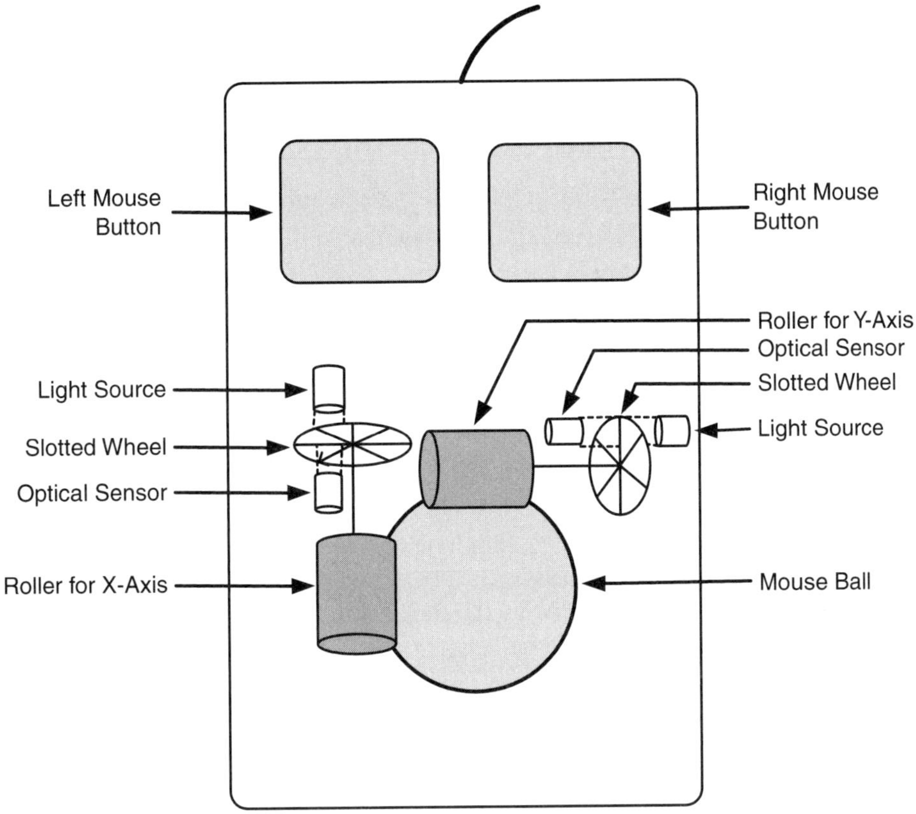

Figure 7.6
Mouse optomechanics.

Actually, the Microsoft Windows operating system and its graphical user interface promoted the mouse to the level it is today. Other devices, such as light pens, graphics tablets, and trackballs, have made an attempt to overthrow the mouse, some with limited success and others with virtually no success at all.

The mouse, used by virtually all PCs, uses an optomechanical design. The mouse ball moves the roller, which is connected to a slotted wheel. The moving wheel allows light to pass through the slots, which are counted by a sensor. The encoding software mouse drivers determine that a movement of approximately 41 slots is the equivalent of ~1 inch of movement on a computer screen (Figure 7.6).

A 6-pin miniconnector connects the mouse to a PC (Table 7.10).

Table 7.10 Mouse Pin Description

Pin Number	Function
1	Data
2	No connection
3	Ground
4	+5 volts
5	Clock
6	No connection

Another pointing device called a trackball is essentially an upside down mouse with a large ball on top. You directly move the ball with your fingers while keeping the device stationary. A trackball works well in a tight workspace.

The touch pad is yet another pointing device; it was originally designed for the laptop computer. The pad senses capacitance from your finger. Movement on the touch pad will be duplicated by the cursor on the screen.

A standard mouse can easily pick up dirt, dust, and grime from your desk, causing it to operate in a jerky or intermittent fashion. It is a good idea to periodically clean the mouse with soap and water or alcohol to prevent this from happening.

One More Look at Personal Computer Modems

If you are not connected through a network (dedicated or virtual dedicated), then a modem may be your only access to the outside world. We've already discussed the basic functionality of modems, but we have not looked at modems specifically for the PC.

The first thing you should know about PC modems is that there are two types: an internal modem and an external modem. Each of the modems performs the same function, but the internal modem uses an internal slot and the computer's power supply, whereas the external modem uses a serial port and an external power supply.

You should also be careful when purchasing an internal modem. All internal modems are the same: NOT. There are software modems (sometimes called WinModems) and hardware modems. A software modem uses the CPU for processing and is less costly than a hardware modem, which has its own processor. As a general rule, it is best to stay away from software modems and spend the extra $50 or so to get a modem with its own processor.

Another interesting fact regarding 56-Kbps modems is their sensitivity to noise and line interference. Also, a power limitation on the local loop imposed by the Federal Communications Commission prevents the modem from achieving its maximum theoretical speed. Generally, 53 Kbps is the maximum speed that can be achieved.

A V.92 standard just released provides the same download speeds as the V.90 (56 Kbps) standard. However, the new standard promises a maximum of 48 Kbps for the upload speed instead of the 33.6 Kbps maximum speed of the V.90. Other features of the V.92 include a "modem on hold" feature that operates with call waiting services and a "quick connect" feature, which improves the time required to connect to an Internet Service Provider.

Before we begin to look at the requirements needed to connect both internal and external devices, let's look at the power requirements to make these devices operational.

Personal Computer Power Supply

The PC power supply converts alternating current wall voltages to low-power direct current voltages. Actually, the PC power supply does much more:

- Upon startup, the power supply must turn on in sequence without any voltage spikes or wide voltage fluctuations.
- Upon shutdown, the power supply must turn off without any voltage spikes or wide voltage fluctuations. This is sometimes called a graceful shutdown.
- Test itself to determine if there is a problem. Upon discovery of a problem, a graceful shutdown should occur.
- Provide its own cooling (usually a fan) to minimize burnout and catastrophic failure.
- Provide a metal cage or steel mesh to minimize electromagnetic interference (EMI). Too much EMI can have a detrimental effect, creating noise that can lead to errors in the data.

Several PC power supply cases were created from original designs made by IBM. The PC power supply most often used is the mini-AT power supply, which provides power output anywhere from 135 to 300 watts.

Transient Voltage Surge Suppression

Transient voltage surge suppression (TVSS) is a low-cost device that should be installed on every sensitive electronic apparatus that is plugged into the utility AC. TVSS protects the PC power supply from transient voltage spikes and incoming noise from the power utility. TVSS can also help prevent permanent equipment damage caused by a lightning strike.

There are several different types of TVSS. Some are whole-house surge suppressors designed to isolate the equipment in the home from the harsh power utility environment outside. Another type of TVSS is designed for each piece of electronic equipment only and may not have an indicator showing when its surge capability has been used up. Always check the manufacturer's specifications to be sure that the TVSS is adequate for your system.

Do not take TVSS lightly. Many catastrophic failures on both the networking (telecommunications) side and the PC side can be attributed to TVSS failure.

A well-designed AC source will have both TVSS protection and uninterruptible power supply (UPS) protection. In many cases, the TVSS is included inside the UPS equipment.

Uninterruptible Power Supply

The uninterruptible power supply (UPS) is one of the best investments you can make for your small business. A UPS is designed to keep full power flowing to the PC in the event of a utility power loss or power brownout. A UPS contains a battery that supplies the power. Since the battery is a DC source, a conversion must take place from DC back to AC. This conversion will provide standard 110 VAC for several minutes or more, depending on the size of the battery and how much wattage is required for backup. This feature is called UPS sizing.

A UPS will also condition the line and provide voltage regulation to supply a pure sine wave for your PC power supply. This is especially important in summer when voltage brownouts can occur.

When the AC wall utility source is good, the power is passed through the UPS and on to the PC power supply. Some of this power from the wall inevitably goes to a charging circuit to keep the UPS battery at full charge. Over time, the battery will lose its ability to take a charge (as will any battery), so a replacement may be necessary.

The UPS must be designed correctly to sense a utility power loss or power brownout and perform a power switchover to its battery. The switchover time should be less than 6 milliseconds.

The UPS output waveform is important also. The desired output from the UPS battery is a pure 110 volt AC sine wave. Some UPS units provide an approximated sine wave by using digitally generated steps or a modified square wave output. You should always buy a unit that has a sine wave output; otherwise, you can damage your PC power supply or your CPU! Check the manufacturer's specifications before making a purchase.

You can size your UPS according to the maximum power load you have plugged into the unit. If you have several PCs, you can obtain an approximate UPS wattage size by adding all of the PC power supply wattage sizes together. Never underdesign your power backup system; always overdesign. An oversized backup system will provide more time for the PC to power down, whereas an undersized backup system may shut down too quickly.

In many cases, the UPS equipment will come with a serial cable and software for your PC. The cable and software are designed to sense a power problem and perform a graceful shutdown before battery power ceases.

Don't forget the power requirements for your video monitor(s). You may wish to connect a low-power flat monitor or no monitor at all to your UPS. Again, the purpose of a UPS is to protect your major equipment investments and to save your data from destruction.

INTERNAL AND EXTERNAL DEVICE CONNECTION

It is always necessary to connect devices together to enable them to communicate. This connection is referred to as an interface. An interface consists of a passageway (usually cable and connectors) by which devices access the CPU.

Two internal interfaces used for mass storage devices such as hard drives, CD-ROM, and removable drives are IDE and SCSI. Both of these interfaces work well. We will examine the advantages and disadvantages of each.

External serial and parallel interfaces are used for modems and printers. We'll examine how each interface performs and what its limitations are.

This section ends with a look at the latest USB and FireWire interfaces. The MIDI music interface and the AGP interface for video are also reviewed.

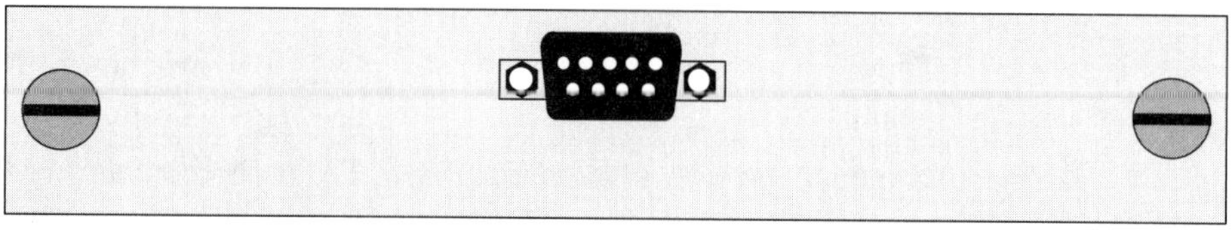

Figure 7.7
Serial port connector.

Serial Port Data Transfer

Serial ports are used to connect devices to a PC and transfer signals one bit at a time. This serial signal transfer of information is in a bidirectional, asynchronous format.

External serial port connectors generally use a DB-9 male connector on the back of the computer case (Figure 7.7).

Each pin on a serial port DB-9 connector has a specific task (Table 7.11).

Serial ports use CPU resources. When a signal arrives through a serial port, an interrupt request (IRQ) is sent. The IRQ signals the CPU that its attention is needed to process the request. Today's PCs use IRQ-3 for COM ports 2 and 4 and IRQ-4 for COM ports 1 and 3. Generally, COM ports 1 and 2 are used for external devices, and COM ports 3 and 4 are used for internal devices.

Table 7.11 Serial Port Pin Description

Pin Number	Function
1	Receive line signal detect
2	Receive data
3	Transmit data
4	DTE ready
5	Signal ground
6	DCE ready
7	Request to send
8	Clear to send
9	Ring indicator

Table 7.12 IRQ Description

Serial Port IRQ	I/O	Port Address
COM 1	4	3F8–3FFh
COM 2	3	2F8–2FFh
COM 3	4	2E8–3EFh
COM 4	3	2E8–2EFh

Since IRQs are shared among internal and external devices, you must always be careful not to cause a resource conflict. For example, if you are installing an internal modem on COM 3, you must be sure not to use the external serial port associated with COM 1. COM 1 and COM 3 both share the same I/O. To avoid resource conflicts, it is best to install external serial devices when their use is not precluded (Table 7.12).

Parallel Port Data Transfer

Parallel ports are also used to connect devices to a PC. However, parallel signals transfer data several bits at a time. Actually, eight pins are used so that an entire byte of data is transferred at one time instead of a single bit of data as in serial data transfer. Therefore, parallel data transfer is considerably faster than serial data transfer. On the negative side, there are some major drawbacks. It is recommended that the maximum length of parallel cable should be no longer than 12 feet to reduce interference and eliminate crosstalk. Generally, serial cable does not have an interference problem.

Also note that with advancements in parallel port signal transfer, it is now possible to have a bidirectional parallel data connection. This type of connection can be used for devices, such as hard drives and cartridge drives, that require a two-way connection. This improvement is sometimes referred to as enhanced parallel port (EPP), which can perform signal transfer rates at speeds from 500 Kb to 2 Mb/sec. If you need to use EPP, be sure to have the appropriate hardware and software requirements needed for IEEE Standard 1284-1994 compliance.

External parallel port connectors generally use a DB-25 female connector on the back of the computer case (Figure 7.8).

Each pin on a parallel port DB-25 connector has a specific task (Table 7.13).

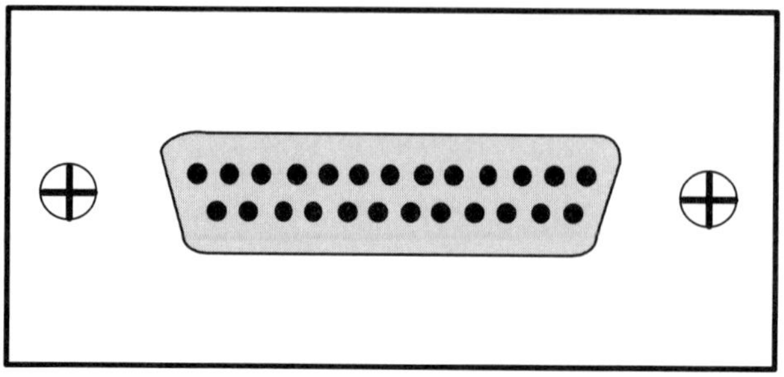

Figure 7.8
Parallel port connector.

Table 7.13 Parallel Port Pin Description

Pin	Function	Signals	Direction
1	Strobe	Internal	Out
2	Data bit 0	Data	Out
3	Data bit 1	Data	Out
4	Data bit 2	Data	Out
5	Data bit 3	Data	Out
6	Data bit 4	Data	Out
7	Data bit 5	Data	Out
8	Data bit 6	Data	Out
9	Data bit 7	Data	Out
10	Acknowledge	Internal	In
11	Busy	Internal	In
12	Paper end	Internal	In
13	Select	Internal	In
14	Autofeed	Internal	Out
15	Error	Internal	In
16	Initialize printer	Internal	Out
17	Select input	Internal	Out

Table 7.13 Parallel Port Pin Description (Continued)

Pin	Function	Signals	Direction
18	Data bit 0	Internal	In
19	Data bit 1	Internal	In
20	Data bit 2	Internal	In
21	Data bit 3	Internal	In
22	Data bit 4	Internal	In
23	Data bit 5	Internal	In
24	Data bit 6	Internal	In
25	Data bit 7	Internal	In

Universal Serial Bus

The universal serial bus (USB) is the next step in PC interface technology. The design of USB is for the connection of low- to medium-speed devices using plug-and-play technology. Plug-and-play enables you to immediately connect or disconnect devices without rebooting the computer.

USB ports operate at speeds of 12 Mb/sec and can connect up to 127 simultaneous devices. Just like serial data transfer, USB provides for a serial connection (1 bit at a time). However, USB transmits its data much faster than standard serial port transmission.

It is expected that USB will replace both serial and parallel ports over time because of its ease of use and simplicity. Virtually all devices are now available in the USB connectivity format. This includes keyboards, mouses, printers, and the like. However, it may be a few years before USB completely replaces serial and parallel connections entirely.

USB ports have four conductors, providing a bus speed of 12 Mb/sec and a subchannel bus speed of 1.5 Mb/sec. The subchannel is intended for low-level communications devices such as a mouse and keyboard. These devices are not intended to consume resources, so the subchannel was created to keep them separate from other devices.

Low-level power can also be obtained from a USB connection. Keyboards and mouses can derive power from USB, whereas cartridge drives, printers, etc., cannot. Higher-level powered devices (over 100 mA) must obtain power from an external power supply.

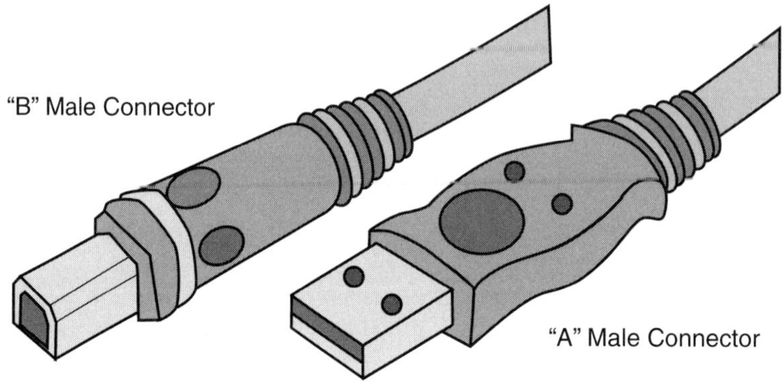

Figure 7.9
USB connectors.

We've already established that USB connections are simple and that they provide moderate communications between the PC and devices. The male "A" side USB connector is rectangular in shape and is called the upstream port. This connector is plugged into the computer. The female "B" side USB connector is a sort of square-shaped connector and is plugged into the device side. The "B" side connector is called a downstream port. There are USB cable extenders with an "A" on one side and a "B" on the other. Try not to use cable with "A" side-to-"A" side connectors. These "A" (male)-to-"A" (female) connector cables can cause problems by shorting out power supplies and rendering devices inoperable!

Some devices contain a hub that enables you to add more devices, or you can purchase a USB hub separately.

USB cable has four wires with a special "shell" connector available on the back of a PC (Figure 7.9).

Each of the pins on a USB connector has a specific task (Table 7.14).

Table 7.14 USB Pin Description

Pin Number	Task
1	Power
2	- Data
3	+ Data
4	Ground

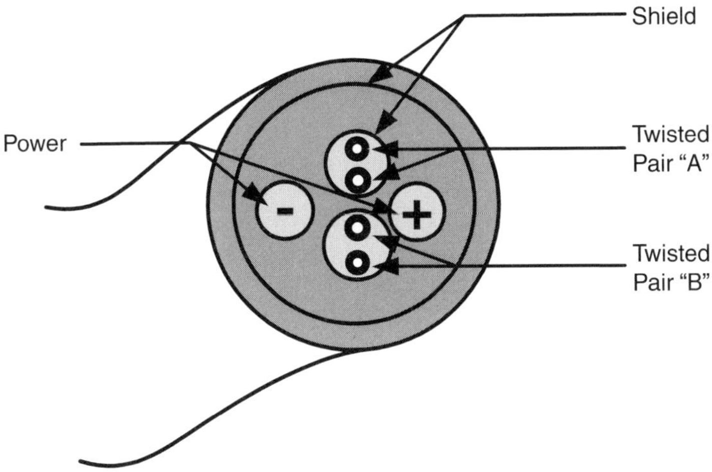

Figure 7.10
FireWire cabling.

FireWire

The IEEE 1394 standard is also known as FireWire. This interface is intended for high-speed devices supporting speeds up to 400 Mb/sec. Devices can also operate at 100 and 200 Mb/sec speeds, supporting up to 63 nodes. Each node can also operate up to 16 devices, so future expansion is available.

Devices such as hard drives, video capture, and camcorders can use a FireWire interface. Many of the newest video devices use FireWire interfaces. FireWire also uses plug-and-play technology, which allows you to add or remove devices without rebooting your PC. Data is transmitted serially, and device power can be obtained directly from the cable in most cases.

FireWire cabling uses three pairs of copper wiring (Figure 7.10).

Each FireWire data pair is shielded to reduce or eliminate interference (Table 7.15).

FireWire can be considered a peer-to-peer bus. In other words, devices can communicate with each other without a computer. An application of FireWire is a camcorder-to-VCR interface without a PC.

Table 7.15 FireWire Pin Description

Number	Comments
1	Signal pair copper (twisted & shielded)
2	Signal pair copper (twisted & shielded)
3	Power pair

Musical Instrument Digital Interface

In a small business environment, you could require the playback of songs for a web page design or business music for the hold file of an answering machine. Sounds might also be needed for advertising or final production work. The MIDI file is a low-cost alternative to music from an original source such as a band or orchestra.

A MIDI file doesn't contain digitized sound recordings as do .WAV (uncompressed audio) or .MP3 (compressed audio) files. Instead, a MIDI file contains instructions on how to create the sound. The instructions are carried out by the sound synthesizer on a sound board containing MIDI instruments. The better the MIDI synthesizer, the more realistic the sound. For example, flute, drums, electric guitar, acoustic guitar, and piano sounds can be recorded separately and combined into one MIDI file. If we play back this combined MIDI file, we would hear the full sounds of a band.

MIDI files use much less disk space than do .WAV or .MP3 files. In fact, many video games and cell phones use MIDI files for the playback of music to capitalize on their minimal space requirements.

There are two types of audio synthesizers for MIDI. The first MIDI synthesizer is frequency modulation (FM) synthesis, and the second is wavetable synthesis. FM synthesis uses artificially created sounds that are modified to sound like musical instruments. Wavetable synthesis uses actual sound samples that are stored in the sound card's memory or wavetable.

You may already realize that wavetable synthesis is superior to FM synthesis because of the actual samples used. However, wavetable synthesis sound cards have varying degrees of quality. High-end sound cards that use 4 MB of memory for the wavetable will have more sounds than a card with 1 MB of memory for the wavetable. With some sound cards, you can include additional memory for the wavetable, allowing more sounds to be permanently stored.

MIDI itself is a serial bus that supports multiple (virtually unlimited) devices using a 5-pin DIN connector (Figure 7.11).

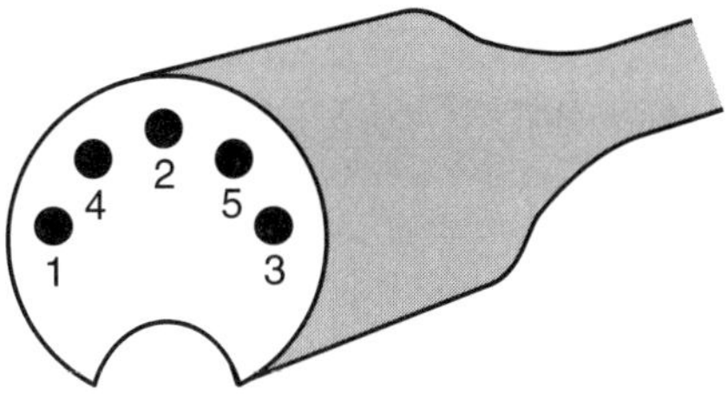

Figure 7.11
MIDI connector.

MIDI devices can be chained together with a maximum cable length between devices of 50 feet.

Industry Standard Architecture

Industry Standard Architecture (ISA) was the first standard in the computer industry developed to support internal devices. Originally, ISA was developed by IBM in an 8-bit bus design. However, with increased CPU speeds, the ISA bus architecture was soon increased to a 16-bit bus design. Internal devices such as modems, sound cards, and older video cards can each use an ISA slot.

Today's PCs still have one or two ISA slots available. It is anticipated that with the expansion of USB and FireWire interfaces, ISA slots will gradually disappear from PC design.

Peripheral Component Interconnect

Peripheral component interconnect (PCI) was originally developed by Intel Corporation as a successor to the ISA architecture previously discussed. Originally, PCI was designed as a 32-bit bus, which was later increased to a 64-bit slot with an additional 32 pins. Gradually, PCI internal devices were standardized on a 32-bit bus.

Internal devices that are equipped for a PCI slot can be configured by software. Another feature of PCI internal devices is their ability to share CPU resources.

Devices such as modems, video cards, and sound cards are all available in the PCI format.

PCMCIA Standards

Personal Computer Memory Card International Association (PCMCIA) is an international standards association founded in 1989 to establish standards for integrated circuit cards used by laptop PCs. Laptop PCs require external devices that are rugged, use low power, and are small.

Standards for PCMCIA have evolved. By 1991, PCMCIA had defined an I/O interface for the same 68-pin connector initially used for memory cards. Socket services specifications and card services specifications were also developed at that time.

There are Type II slots and Type III slots for PCMCIA. Type II slots can accommodate memory cards, network interface cards, and modem cards. Type III slots (which are two Type II slots stacked one on top of the other) can accommodate a larger (mini) hard drive.

Generally, PCMCIA slots are available only on laptops, but some adapter cards are available that allow you to use a PCI slot on a desktop computer.

Accelerated Graphics Port

The accelerated graphics port (AGP) is a dedicated port designed exclusively for video. The key factor regarding AGP is its fast video-processing speed. The AGP slot operates at the speed of the CPU bus. For example, a PC with a CPU bus speed of 66 MHz and containing an AGP slot can operate at video port speeds of 66 MHz. In addition to the increased speeds, with AGP, main CPU memory is accessed directly for the complicated process of texture mapping. PCI video cards can also provide texture mapping, but it is a slower process because main CPU memory cannot be directly accessed.

Without getting too technical, we can say that AGP is an effort by computer manufacturers to provide a bus speed that is virtually equal to the CPU bus speed and that allows for faster video processing. If you have a choice, you should always use an AGP video card as opposed to a PCI video card.

Integrated Drive Electronics

Integrated drive electronics (IDE) performs a hard drive controller function for the connection of a hard drive into the computer's CPU bus. Originally, an IDE hard drive contained a separate hard drive and an IDE controller card with a 40-pin ribbon cable connection. Today, the IDE hard drive controller is standard on all PC motherboards, with primary and secondary connectors accepting up to four internal devices (such as IDE CD-ROM drives and IDE hard drives).

Almost all IDE hard drives are configurable with jumpers or switches, as follows:

- Master drive (single drive)
- Master drive (dual drive)
- Slave or secondary drive (dual drive)
- Cable select (dual drive-special cable)

Generally, it is best to keep drives separated by using either the primary or secondary channel. A CD-ROM device on the same channel as a hard drive will ultimately slow the hard drive's performance.

Small Computer Systems Interface

Small computer systems interface (SCSI) is designed as an interface for hard drive and CD-ROM devices. SCSI is usually installed in higher-quality (and more expensive) PCs supporting more devices (such as SCSI scanners and SCSI tape drives). Up to seven devices can be on one link.

SCSI uses parallel communications between a host adapter and its devices. Either 8 or 16 wires are used to transmit data. Therefore, wires must be short enough to prevent crosstalk and interference.

A SCSI interface is generally not installed on a motherboard (the IDE interface is almost always on a motherboard). Therefore, a SCSI host adapter must be installed in a PCI slot. SCSI devices can be connected to the host adapter either internally or externally.

SCSI internal devices use a 50-pin ribbon cable, and external devices use two connectors. One external connector can connect to the host adapter while the other connector extends the SCSI interface to the next device.

Many types of SCSI interfaces are available: narrow SCSI and wide SCSI configurations and several variations of each.

Table 7.16 SCSI Description

Connector	Description	Speeds
SCSI-1	Uses Centronics 50, Micro-Ribbon 50, or Telco 50 connector	5 MB/8-bit bus
SCSI-2	Uses Micro-D 50 pin with thumb clips or Ribbon 60 connector	10 MB/16-bit bus or a Micro-40 MB/32-bit bus (wide)
SCSI-3	Uses Micro-D 68-pin connector	20 MB/16-bit bus

Each device on the SCSI bus must have a SCSI ID number. The device number on each link must be unique. The most common cause of SCSI problems is two or more devices sharing the same SCSI ID. The highest-numbered device has the highest priority on the SCSI chain; the lowest-numbered device has the lowest priority. The IDs are controllable by the user, so in most cases, the SCSI hard drive would be assigned the highest number.

Another potential problem with the SCSI connection is its termination. A termination resistor must be installed after the last device on the chain to signify the end of the chain. Many SCSI devices have switchable termination resistors that will signify that a particular SCSI device is at the end of the chain. An improperly terminated SCSI chain can cause a device conflict and possibly can damage equipment! Read the manufacturer's instructions for each SCSI device before attempting to boot up the PC.

To complicate the issue, there are many types of SCSI configurations. SCSI-1, SCSI-2, and SCSI-3 devices all operate at different speeds and use different connectors. Make sure that you have the proper cables, host adapter, and devices before attempting an installation (Table 7.16).

The development of SCSI-2, or fast SCSI format, provides for faster speeds on the overall bus. Wide SCSI increases these speeds even further. Wide SCSI uses an extra twisted pair, which doubles or quadruples the speed over the bus.

The newest SCSI, Ultra-2 low voltage differential (LVD), provides increased speeds up to 80 Mbps and cable lengths up to 39 feet! SCSI specifications have changed considerably since their inception in the early 1980s (Table 7.17).

Table 7.17 SCSI Requirements

Type	Bit Width	Mbps	No. of Devices
SCSI-1	8	5	8
Fast SCSI (SCSI-2)	8	10	8
Fast Wide SCSI (SCSI-2)	16	20	16
Ultra SCSI	8	20	8
Ultra Wide SCSI (SCSI-3)	16	40	16
Ultra 2 SCSI (LVD)	8	40	8
Ultra Wide 2 SCSI (LVD)	16	80	16
Ultra Wide 3 SCSI (LVD)	16	160	16

So which type of hard drive interface should you use? The general rule is that IDE should be kept for the PC workstation because of its low cost, simplicity, and ease of installation. SCSI devices generally outperform IDE devices and should be used for PC servers because of their superior speed.

PERSONAL COMPUTER OUTPUT DEVICES ....

We cannot complete our discussion of PCs without first looking at external output devices. These devices are well worth the extra investment since they may outlast other PC components, including hard drives, CPUs, and memory, that become outdated quite frequently.

Cathode Ray Tube Video Monitors

The cathode ray tube (CRT) still remains the best choice for PC video display. Today's CRTs are smaller, lighter, and inexpensive. The CRT for the PC functions exactly the same as a CRT for television, which we covered in Chapter 4.

The standard size PC monitor in use today is 17 inches, with a display of 1024 x 768 dots per inch. Nineteen-inch and 21-inch monitors are also popular, but the costs are much higher.

One good indication of a video monitor's quality is its dot pitch specification. Dot pitch refers to the distance between diagonally adjacent dots of the same color. Most CRTs display colors in an arrangement of red, blue, and green dots. The three colored dots are set in a triad or triangular pattern. Ideally, the dot pitch should be anywhere from .22 to .29 mm, with the lowest number producing the highest quality.

We already discussed refresh rate in this chapter. A monitor should be capable of producing high enough refresh rates to avoid video flicker.

One highly important factor concerning monitor performance is the controls that determine how well the video image is perceived. Another factor is the power consumption. Monitors consume higher levels of power, so you want to be sure that the monitor you choose has advanced configuration and power interface (ACPI). With this feature, the monitor consumes lower levels of power in periods of user inactivity.

Liquid Crystal Display Video Monitors

One of the most popular (and expensive) devices to flood the PC market in recent years is the liquid crystal display (LCD). These video display devices consume low power, are not subject to electromagnetic interference, conserve space, and are relatively lightweight. Until recently, LCD displays were exclusive to laptop PCs. However, with improvements in design and the reduction of cost, LCDs are competing directly with CRTs for the PC video display market.

LCD video displays contain polarizing filters arranged at 90° angles with liquid crystals between them. The crystals twist the light passing through them toward the first filter and then to the second filter, which in turn illuminates the screen. An electrical current is then applied to the crystal, forcing a reset. The process of a lit LCD display followed by an electric current/crystal refresh is what changes the image on the screen.

Early LCD monitors used a passive matrix display. This type of display provides current as a series of pulses to an entire row or column of pixels. The transistors, required to produce an electrical current, are kept along the horizontal and vertical edge of the display. Therefore, a passive matrix 1024 x 768 display would contain 1,792 transistors.

Passive matrix displays have several disadvantages, however. Smearing and streaking of colors on the display, along with lack of brightness and limited viewing angles, contribute to their relatively poor quality. The biggest advantage of this type of display is its relatively low cost and lightweight capability.

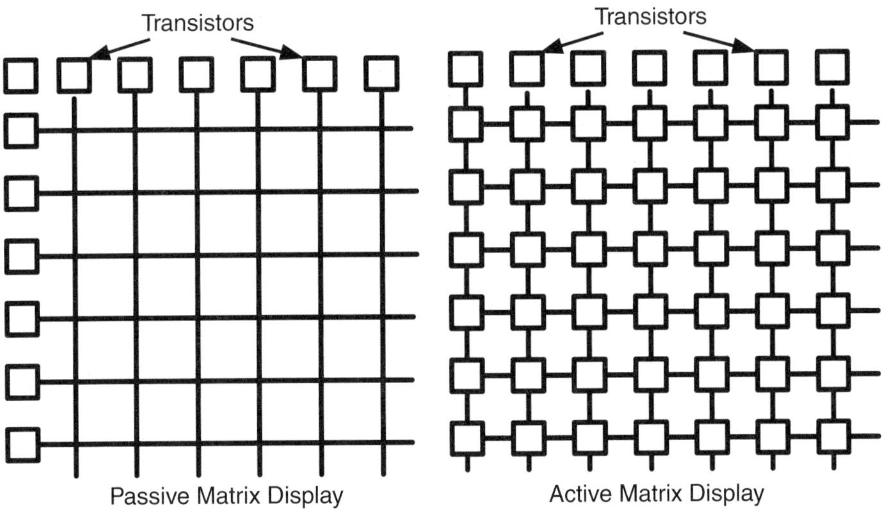

Figure 7.12
Active vs. passive matrix display.

Active matrix displays contain many more transistors than do passive matrix displays. In fact, active matrix displays have a transistor dedicated to each pixel. With so many transistors required for this display, the transistors must be located directly behind the viewing screen. This type of display is called thin film transistor (TFT). There is no need to pulse the current because each transistor is dedicated to its own pixel. For example, a 1024 x 768 display would have 2,359,296 (one for each color) transistors instead of the 1,792 transistors required for an equivalent passive matrix display (Figure 7.12).

Because each transistor can control the amount of light passing through the filters with much more accuracy, a brighter picture is obtained. Smearing and streaking are virtually eliminated, and the overall quality is vastly improved over passive matrix displays.

If you are looking to purchase an LCD with active matrix, you should note that some screens will inevitably have failed transistors, causing pixels on the screen to remain permanently light, dark, or somewhere in between. Manufacturers generally do not guarantee 100% accuracy with their active matrix screens, so it is best to check the display before purchasing the product.

Printer Technology

Printer technology uses dots per inch (DPI) to represent the quality of resolution. DPI refers to the number of dots required to create a line one inch long. The more dots displayed in a given area, the better the quality of the image. For example, a 600 x 1200 DPI printer contains 720,000 dots to create a square inch. In this case, there are half as many horizontal dots as there are vertical dots.

A 300 x 300 DPI printer is considered a low-resolution printer (by today's standards). Characters contain jagged edges because the dots are unable to fill the gaps. With increased resolution and DPI, you cannot notice the jagged edges as easily but upon close examination you will see that they are still there.

Remember that for each dot created (in black), one bit of memory is used. So the higher the resolution, the more memory required.

Some manufacturers resort to resolution enhancement by interpolation. Interpolation is a method for faking a higher resolution by adding dots to fill in the spaces, resulting in an enhanced resolution. Unfortunately, printers that use interpolation are not as good as printers with a full DPI. For example, one printer is advertised with a 600 x 1200 DPI, and another printer is also advertised with a 600 x 1200 DPI with interpolation. Although the first printer may be more expensive, the quality is improved over the second one. Keep in mind when shopping that DPI rates aren't the full story.

It is important to note that both laser and inkjet printers have resolutions far exceeding standard video monitor resolutions, which average somewhere around 70 DPI!

Before we take a closer look at the more popular laser and inkjet printers, let's focus on that old workhorse called the dot-matrix printer.

Dot-Matrix Printers

Until several years ago, dot-matrix printers were the standard for office printers. They were low cost and did an adequate job. The printers themselves worked much like a typewriter, using dot strikers instead of a single hammer striker. Laser and inkjet printers have replaced most of the dot matrix printers. In certain cases, however, these printers still have a useful function.

The print head itself consists of a matrix of pins, usually 24 pins. Each pin corresponds to a dot, which in this case is a 4 x 6 DPI. The appropriate pins for each letter or number are pushed out and strike the inked ribbon. The ribbon is then pushed against the paper, forming the corresponding letter or number from the pushed-out pins.

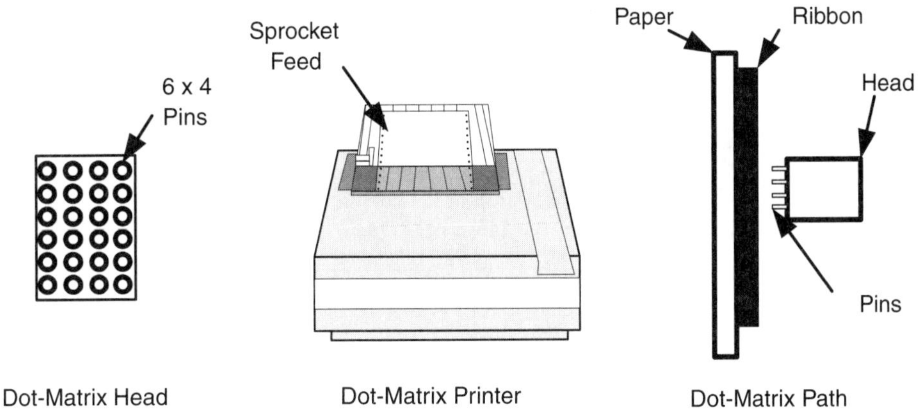

Figure 7.13
Dot-matrix printing.

Compared to other printer technologies, the characters are large and cannot be made smaller. New dot-matrix printers are in limited supply and may cost more than other kinds of printers.

I still use dot-matrix printers to print checks and to print invoices on invoice forms. The printers are reliable and low cost, and they perform the job as long as images are not needed. These printers work especially well with sprocket feed paper. This paper has perforated holes on each side to correspond to gears on each side of the printer. The gears pull the paper through the printer as needed (Figure 7.13).

Laser Printers

Laser printers provide the best value for the dollar. Prices have dropped considerably and the quality is outstanding. The current standards for laser printers are 600 x 600 DPI with high-end commercial printers in the 2400 x 2400 DPI range.

These printers operate with a photosensitive drum that contains electrically charged areas corresponding to darker images. An ultrafine black powder, which is usually plastic, is heated and rolled over the drum, adhering to the electrostatically charged particles on the drum. The drum then rolls onto a sheet of paper and deposits toner, keeping the same pattern as the original image. Finally, the paper is heated, which fuses the toner to the paper, resulting in a final copied image (Figure 7.14).

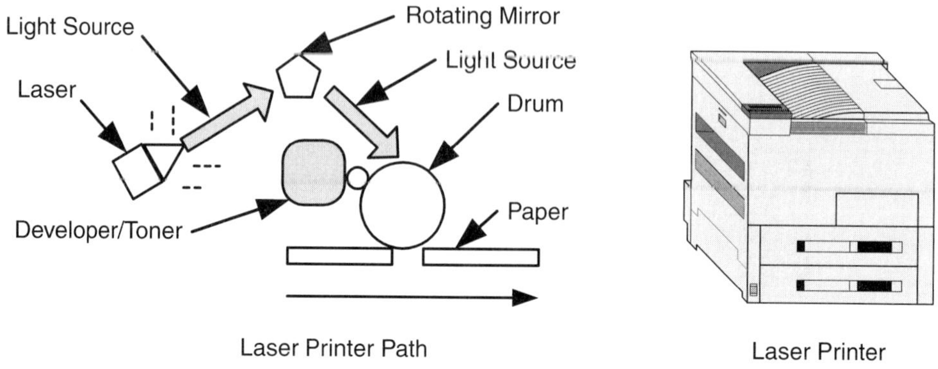

Figure 7.14
Laser printer technology.

The process is similar to the process used for photocopier machines. The main difference is that a photocopier must scan the image first, whereas a laser printer is fed the image from a PC's output parallel port (in most cases). Some laser printers are multifunctional. These devices have a page scanner, a photocopier, and a printer all in one.

Periodically, you must replace the toner because it is consumed and deposited on paper by the laser printing process. Manufacturers tend to package toner in sealed cartridges since the toner itself behaves like ink and can stain very easily. Some manufacturers now provide a cartridge containing a photosensitive drum and developer/toner in one unit. This can be a fairly expensive replacement, and it is important to note that the photosensitive drum needs periodic maintenance. However, an all-in-one cartridge combining the drum, toner, and developer actually ensures good performance with a minimum of downtime.

To save money, some vendors sell refill kits for your laser printer's all-in-one cartridge. A refill requires you to punch in a refill hole and then seal it up. Again, unless saving money is your only requirement, it is still best to follow the manufacturer's recommendations and replace the entire assembly.

Inkjet Printers

The inkjet printer is a simple yet effective printing device for the small business environment. The printer itself produces a decent document but cannot provide the quality of laser printer output. Therefore, the unwritten rule is that home users buy inkjets and business users buy laser printers.

Inkjet printers tend to be much less expensive than laser printers. You can purchase an inkjet for as little as $50. Inkjets do not contain a drum or developer and use liquid ink instead of toner. The ink is contained within a cartridge. The cartridge has nozzles that spray the ink on paper in much the same way that a can of spray paint works. These nozzles are extremely small as compared to a can of spray paint. Sometimes 128 nozzles or more are used per color.

Nozzles over time can gum up and dry out, so manufacturers tend to provide a cleaning software program that clears the nozzles. This cleaning procedure restores the printer output to its best possible quality.

Inkjet printers cannot produce detail as fine as that which a laser printer produces. Moreover, inkjets tend to smudge and smear on certain types of paper. Therefore, paper quality is always an issue with inkjet printers. You should purchase paper that is recommended for inkjets. This paper is usually more expensive than standard paper stock.

There are basically two types of inkjet printers. One operates thermally (with heat); the other uses crystals that react to an electric current.

The Bubblejet printer, pioneered by Canon Computer Systems Inc., uses a heating element that boils the ink to ~400°F. The boiling ink creates a gas, and pressure forces ink through the nozzle. This type of printer is also known as a thermal inkjet printer.

Epson Corporation was the first company to produce an inkjet printer that used piezo crystals in each individual nozzle. The crystals react to an electric current by changing shape, which in turn forces the ink out and onto the paper. Piezo ink cartridges tend to last longer than thermal ink cartridges because high heat is not a factor.

Virtually all inkjet printers manufactured today can produce color output. It is possible to create all colors by mixing cyan, magenta, yellow, and black in different proportions. This is known as CMYK printing, where the "K" represents black. Several recent improvements in color inkjets have produced extremely good color output for printing images or photos. Be sure to have photo-quality paper when attempting to reproduce a photo on an inkjet printer.

Label-maker

There is one more printer I would like to briefly discuss. It's called the label-maker. The label-maker has a specific task. You guessed it: to print labels. Thermal print technology burns the dots onto the plastic or paper label, with a 300-DPI resolution. This DPI is more than adequate for label printing. Many of these printers can create a

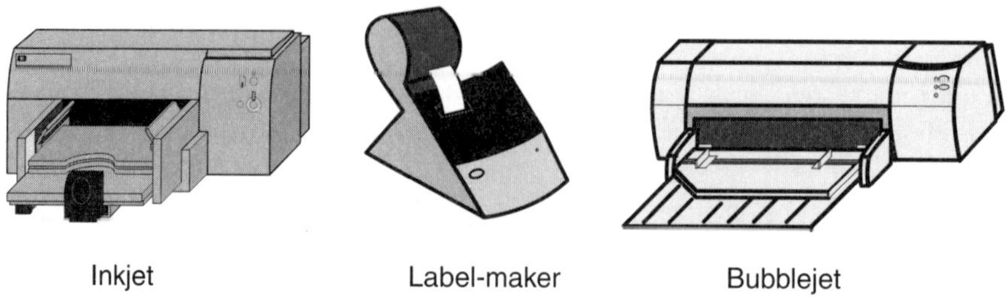

Figure 7.15
Inkjet, label-maker, and bubblejet printers.

label in as little as two seconds and also come with the software needed to create bar codes and generate addresses from email contact lists.

Label-makers are much smaller than standard printers and can be very useful. These printers can be connected with a USB or a serial port (Figure 7.15)

PERSONAL COMPUTER NETWORKING..................

What would the world be like without networking and telecommunications? PC development has grown substantially in the first 20 or so years of its existence. What would the PCs be like if we couldn't connect them together? How would we transmit or receive the information that can be so vital to our business?

We discussed client/server computing in Chapter 5. Client/server technologies share the processing requirement between both client and server. In many cases, the client is the desktop PC and the server is the web server. The web server handles the request for information from the client and transmits the information back to the web browser and from the browser back to the client. This application is called the wide area network (WAN).

In an office environment we would like to share information locally between co-workers, management, administrators, etc. This is where a local area network (LAN) will be required.

Local Area Network

A LAN is required to link a relatively small number of desktop computers in an office environment. There are two types of connections you can create for a LAN environment: you can allow only one transmission at a time for the entire LAN; or, knowing that performance may suffer, you can allow anyone to transmit anytime.

A LAN can be in a star, mesh, bus, or ring configuration (see Chapter 3). The star configuration with Ethernet protocol, however, is generally the standard configuration used.

Network Adapter

The network adapter or network interface card (NIC), connects a PC to the local network. Today, many PCs have their NIC integrated on the motherboard. Laptop PCs can use a PCMCIA device card for the NIC. Note that some PCI and ISA NIC cards have local network access capabilities.

An NIC can be purchased to operate with Ethernet protocol or Token Ring protocol. The NIC is generally available for unshielded twisted pair (UTP) connections. Adapters can be purchased to convert a UTP RJ-45 8-pin connector to a coaxial connection.

Ethernet Protocol

The most popular networking protocol for the LAN is Ethernet. The Ethernet standard was originally developed in the 1970s by Xerox Corporation and used coaxial cable. Today, most Ethernet cabling uses UTP (Table 7.18).

Table 7.18 Ethernet Description

Name	Designator	Max. Length	Cable Type	Connector
Thick Ethernet	10Base 5	500 meters	RG-8U	N
Thin Ethernet	10Base2	185 meters	RG-58—U	BNC
Twisted-Pair Ethernet	10BaseT	100 meters	Cat.3 UTP	RJ-45
Fiber-Optic Ethernet	10BaseFl	1000 meters	62.5 fiber	ST

Standards have been revised, and Fast-Ethernet can now run on UTP at 100 Mbps (Table 7.19). Future applications call for Fast-Ethernet on UTP running at 1 Gbps (see Chapter 10).

Table 7.19 Ethernet UTP Description

Designator	Max. Length	Cable Type	Connector
100BaseTX	100 meters	Cat.5 UTP (2-wire pair)	RJ-45
100BaseT4	100 meters	Cat.3 UTP (4-wire pair)	RJ-45
100BaseFX	412 meters	62.5 multimode fiber	ST

Ethernet protocol contains carrier sense multiple access with collision detection (CSMA/CD). CSMA/CD allows each PC equal access to the network. The problem, however, is that each workstation cannot transmit at the same time. We want to avoid any simultaneous transmissions or data collisions if possible.

When a PC needs to transmit data, it first listens to the local network to see if anyone else is sending. This is the carrier sense multiple access (CSMA) part of Ethernet. If detection shows that the local network is busy, the PC waits and then rechecks. This rechecking is done again and again until the local network is clear of any data. Once the local network is clear, the PC can begin transmission.

It is still possible, however, for two simultaneous local network transmissions to occur, effecting a collision. Usually this happens when one PC has begun to transmit and the signals have not yet reached the other PC, which also wants to transmit. This is the collision detect (CD) part of Ethernet.

When a collision occurs, both PCs back off, pause for a short period of time, and then one of them begins to transmit. If a collision with the same transmission occurs again, Ethernet protocol will require a longer pausing time and the process starts over again.

Collisions are not necessarily a bad thing on an Ethernet network. Usually, on a properly designed local network, the number of collisions is small and the network is not affected overall.

UTP Ethernet connections use an RJ-45 jack. Only one pair is used for a 10 mbps circuit while two pairs are used for a 100 mbps circuit (Figure 7.16).

If the local network has grown too large, collisions occur more often, and efficiency is down, you may want to consider using a bridge, router, or switch. We take a closer look at these devices after we discuss Token Ring protocol.

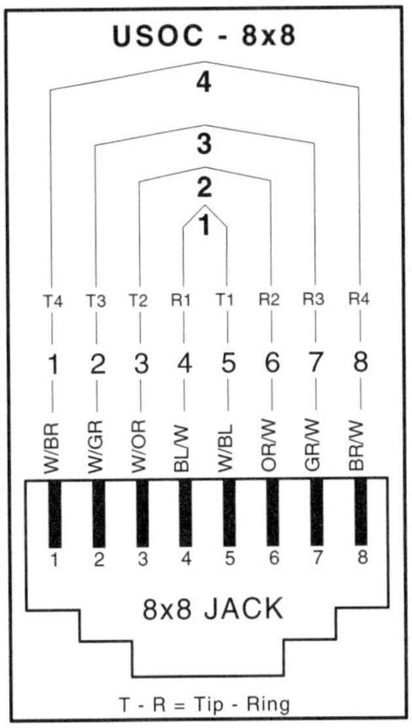

Figure 7.16
Ethernet connector.

Token Ring Protocol

The other LAN protocol is Token Ring. Token Ring was initially developed by IBM in the early 1980s. Token Ring is more complex than Ethernet. Equipment for Token Ring is more expensive and the maximum speed is 16 Mbps.

Token Ring uses a ring topology whereby a single token is passed around the ring from one PC to another PC. When a PC wants to transmit, it captures the token from the local network and adds its data to the packet with a destination address, and the token continues around the ring. Similarly, a token that contains a data packet is sent around the ring until it reaches its final destination.

So the token carries one datum at a time to its final destination. There are no collisions on this type of network, so the local network performs well under high-volume conditions. However, Token Ring's limitations of speed and higher cost have kept this technology well behind Ethernet in popularity.

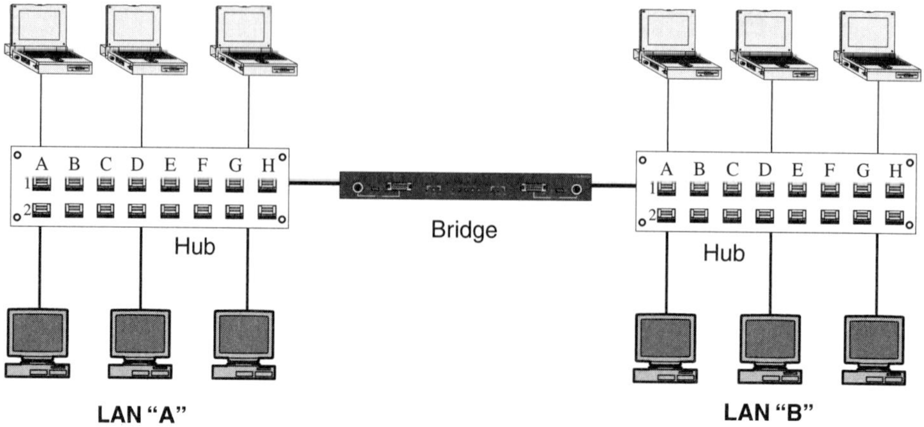

Figure 7.17
Bridge.

Bridges

A bridge connects two LAN segments. This device reads the destination address of the packet and determines whether the packet needs to be sent over the bridge to the other LAN segment or can remain in the same LAN segment. This approach eliminates or reduces unnecessary data from traveling to the other LAN segment, thereby increasing efficiency.

It is necessary to use a bridge when your LAN segment becomes too large and collisions are occurring too frequently. Another example is the creation of a separate LAN segment for your sales and marketing group and another LAN segment for your product development and engineering group (Figure 7.17).

Ethernet uses a process, called transport bridging, which builds its own tables automatically. Tables are needed to determine which side of the bridge each address is located on. If a bridge does not have this automated capability, then you must manually enter the routing address information into the tables.

Routers

Routers can also be used to separate LANs into two segments. Although bridges have some filtering capability, they still permit broadcast transmissions throughout the

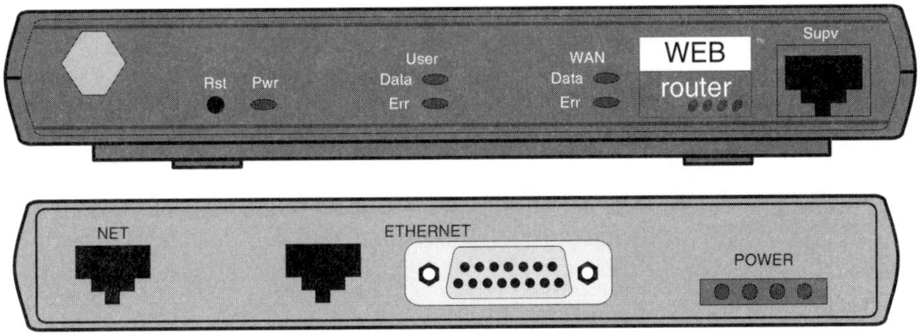

Figure 7.18
Routers.

entire network. Routers have filtering capability and can perform other functions such as connecting networks with dissimilar protocols (i.e., Ethernet and Token Ring). We already know that routers are intelligent devices that determine the most efficient route for Internet Protocol (IP) traffic.

Remember that routers must learn from their environment. There is no master list of all networks or all routers. Instead, routers must learn about their surrounding networks in their routing tables and share their information with other routers (Figure 7.18).

Some routers, called specialized routers, are specifically designed for local access. For example, a router can be connected to an xDSL or cable modem. This router would allow multiple PCs to share broadband access. Network address translation is used with these local routers to help prevent security threats from attacking the network; in that way, the router behaves as a firewall.

Hubs

Hubs, or multiport repeaters, are used to create a star configuration in a local network environment. A hub also contains an amplifier or repeater function that allows the cable for each PC to extend up to 100 meters long.

Some hubs have an uplink port, which enables you to extend hubs in a daisy-chain fashion from one to the other.

Minihubs are available in 4- or 8-port versions. Standard hubs generally come with 8, 16, 24, or 32 ports (Figure 7.19).

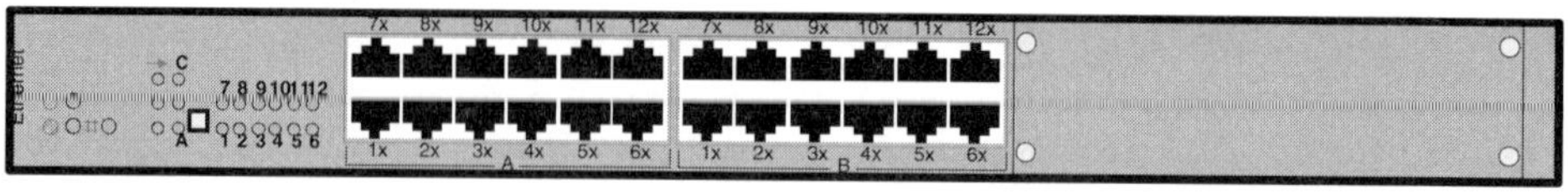

Figure 7.19
Hub.

Switches

A switch is an intelligent hub. The switch reads the contents of a packet and then transmits the packet only to the port or ports that require the packet. This behavior differs from that of a hub, which transmits the data to all of its ports.

Local switching is desirable since each packet is sent only to its destination port, creating a dedicated segment between the originating PC and the terminating PC or network. With a dedicated segment, packet collisions cannot occur and each connection has the full bandwidth of the network available.

Wireless Local Networks

The wireless networking standard is the latest phase in the evolution of LANs. The network is created through unguided radio signals instead of traditional guided copper wires. The installation of a wireless network can be easy compared to installation of a wired network. Rearranging the office can be a snap with a wireless network compared to a wired one. Installations are completely portable within a building or office complex. There are some drawbacks, however, as we shall soon see.

The original wireless network specification was developed by the Institute of Electrical and Electronics Engineers (IEEE) and called IEEE 802.11. This original specification operated at speeds up to 2 Mbps in the 2.4 GHz range, using spread spectrum (SS) transmission technology.

SS technology allows data in small chunks to be sent over several discrete frequencies in a specified range. In other words, the operating transmission frequency is always changing and "spread" among several frequencies. Basically, wireless networking technologies using spread spectrum can operate in a direct sequence spectrum (DSS) mode or a frequency hopping spread spectrum (FHSS) mode. We'll discuss more about spread spectrum in Chapter 9.

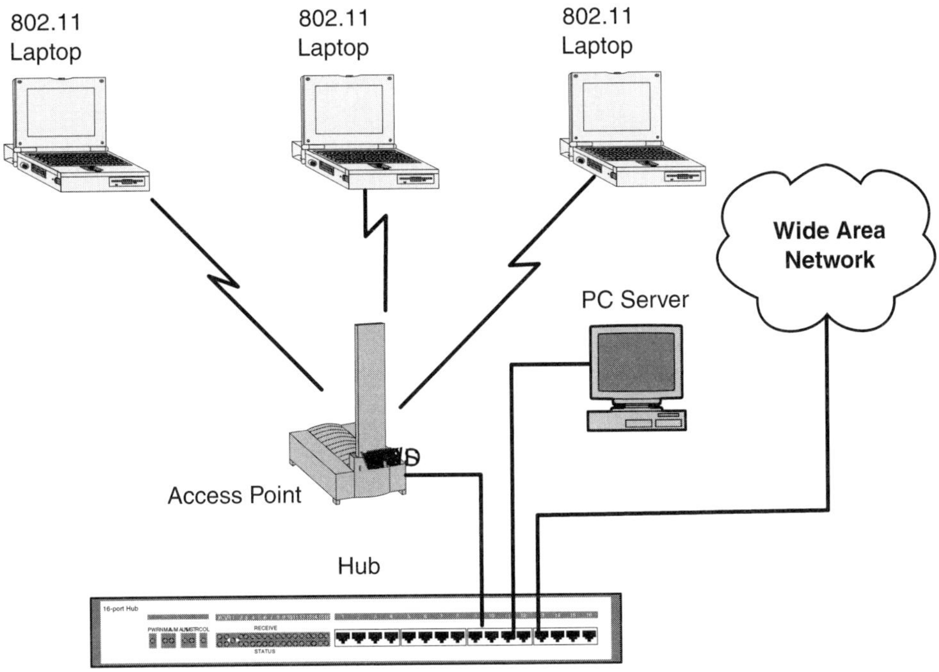

Figure 7.20
Wireless local network.

DSS transmission operates by splitting each byte of data into several parts and sending them simultaneously on different frequencies. FHSS transmission operates by sending a short burst of data, changing the transmission frequency, and then sending another short burst of data. The transmission time on each frequency should last no more than 400 milliseconds. Both the transmitting and receiving equipment must agree, based on algorithms, to the transmission frequency to be used. FHSS devices are less costly than DSS devices but are not as fast.

If you have a laptop PC, a wireless network may be ideal. A wireless network PCMCIA card can be used in a laptop's PCMCIA slot. PCMCIA will allow you to remain connected to the network as the PC moves from room to room. The base unit/access point device of a wireless system behaves like a hub connecting to the network and providing wire-free access to all laptop PCs that are configured for the network (Figure 7.20).

Another revision to the IEEE 802.11 is IEEE 802.11b, also known as wireless fidelity (WI-FI). This revision uses DSS and can run at a relatively fast speed of 11 Mbps. Poor signal strength or increased interference will drop the speed to as little as 1 Mbps but will keep the wireless connection running.

WI-FI is somewhat expensive. A base station/access point can cost from $500 to $1,000, with each PCMCIA card costing from $150 to $300. However, depending upon the situation, the benefits of wireless PC workstations may far outweigh the negatives in cost and setup time.

Here is a word of caution, however, about security. If the base station or access point is not coded correctly with wireless network security, it is possible to obtain easy access to your network. The Wired Equivalent Privacy (WEP) standard provides privacy for WI-FI.

A lack of security on a wireless network is a boon for hackers who have been known to drive by major corporations in a car and obtain access to major corporate networks. They succeed in obtaining network access by using a laptop running the IEEE 802.11b standard.

Phone Line Networking

Phone line networking is based on specifications developed by the Home Phone Networking Association (HPNA). The network is relatively inexpensive and easy to install; however, you cannot use existing wiring in your home or office. Phone line networking does combine both standard analog and data services on one line, eliminating the need for a second wire.

The line length limit is 1000 feet, and the overall coverage area is about 10,000 square feet. Note that the maximum speed of this type of network is 10 Mbps (version 2.0).

Frequency division multiplexing (FDM) is used to separate voice signals from data signals, and there is plenty of bandwidth on the wire to carry both.

Sometimes a low-pass filter is needed to isolate the voice and data signals to better keep interference to a minimum. Phone line networks should also use data encryption.

Power Line Networking

Power line networking uses the 120 VAC power lines that are present in your home or office. This technology is called PowerPacket. Every room has at least one AC outlet, so the cost for wiring is virtually zero. The cost for adapters is expected to be minimal.

This new technology is relatively fast, operating at 14 Mbps. Oh yes, this technology also avoids power line brownouts and blackouts by maintaining its speed when these conditions exist.

PowerPacket technology uses orthogonal frequency division multiplexing (OFDM), which separates available frequencies on the power line into 84 channels. If interference, a power surge, or noise occurs on the line, the PowerPacket circuit compensates; it relies on certain channels (more than others). The power line technology should remind you of rate adaptive DSL technology, which we've already reviewed.

The future of power line networking technology is expected to be very bright for the home and small business user.

CONCLUSIONS ....

This chapter presented several important topics. The PC was opened and we took a look inside. Internal components of the PC, such as the CD-ROM, hard drive, and CPU, were examined in some detail. Because these technologies are so pervasive in our society, it is important to understand how they operate.

We reviewed interfaces to both internal and external components and became familiar with the differences between SCSI and IDE, USB and FireWire, and ISA and PCI.

PC output devices such as the solid state video screen and CRT are important, especially for graphics and video. We looked at the various types of printer technologies such as dot-matrix, laser, and inkjet. We discovered that the laser printer is well suited for the office environment, whereas the inkjet may be better suited for home use because of its color and graphics capability.

At the end of this chapter, we started to inspect our local data network protocols, covering Ethernet and Token Ring. Local network equipment such as routers, switches, and hubs were defined. We concluded with some of the newest technologies that use power line and phone line technologies.

In the next chapter we will look at software requirements for the small business environment.

8 Personal Computing Software and Security

Now that we have some familiarity with PC hardware, we need to look at another important aspect of computing: PC software. Software provides the instructions on what the PC needs to do and how to do it. Software enables humans to interface with the PC. We provide information input; the PC processes the information and provides information output in a desirable format.

Software has become big business, and Microsoft has the lion's share of it. There are programs to produce documents, store contact information, create and read email, play music, watch videos, edit diagrams, edit photos, create spreadsheets, browse the WWW, etc. As hardware improves, so does the software to coincide with the changes in hardware.

This chapter is by no means an in-depth study of software requirements for the PC. Instead, I've chosen to outline the most important aspects of software needed to keep your PC running efficiently, error free, securely, and virus free.

It is important to become familiar with software upgrades. These upgrades can resolve potential problems and provide new and enhanced features. Many software developers provide free downloads, email notification, and automated detection and correction.

Some of the biggest uses for the PC are email and the World Wide Web (WWW) browser. Because of the WAN and the use of IP, we can now share and distribute data with relative ease. However, we can also share and distribute viruses and spam email, which we discuss in this chapter with some detail.

If I want to know what is going on at the office when I'm on vacation, I'll check my email first before going to voice mail (answering machine). The reason is that society and the business world have become so comfortable with email and web browsing that many times a voice message is not even necessary. Email provides you with a documented response: "something in writing," as they say.

Security in a small business or consumer environment is sometimes overlooked. An invasion of privacy or inadequate security can be detrimental. Lately, there has been considerable talk about stolen identity and data management, so we look at what it takes to keep the security risk factors to a minimum.

The first section of the chapter takes a brief look at PC operating systems. All PCs need an operating system to process software programs needed to perform a task.

The second section covers applications software. Today, we have applications software designed for specific tasks such as office, multimedia, and photography environments, just to name a few.

The third section examines how to save data, back up data, and keep data safe and secure. I cannot overemphasize the importance of saving and securing data correctly.

Section four is concerned with email applications—privacy and security—and section five looks at browser applications: privacy and security.

Section six covers the all too familiar problem of virus attacks on your computer. Staying virus knowledgeable and keeping a watchful eye is the key to keeping these attacks to a minimum.

Taaa.Daaa..

OPERATING SYSTEMS

We need an operating system (OS) to translate between the CPU, its associated hardware, and the software itself. Without this translation, we would need customized applications software designed for every type of PC hardware sold on the market today. Thus, the OS is the interface between the hardware and the applications programs.

Operating systems have improved vastly in the last several years. Point-and-click OS's have provided even the novice user with the ability to perform functions that were once available for high-end sophisticated users with disk operating system (DOS) commands.

Microsoft's Windows is by far the most popular OS available today, so I'll concentrate on the differences between each of the Windows OS environments. Others, such as Apple and Linux, are also briefly discussed.

Timetable of Historic Windows OS Events

1983	At the Plaza Hotel in New York, Microsoft formally announces Microsoft Windows.
1984	Microsoft releases MS-DOS 3.0 for PCs, providing support for 1.2 MB floppy disks, and 10 Mbps hard drives.
1985	Microsoft ships Microsoft Windows 1.0, for US$100.
1987	IBM and Microsoft announce Operating System/2 (OS/2) for IBM PS/2 PCs.
1988	IBM ships DOS 4.0 containing support for 32 Mbps hard disk.
1989	Microsoft introduces and ships Microsoft Windows 3.0.
1991	Microsoft releases MS-DOS 5.0 containing a full-screen editor and additional utilities.
1992	Microsoft ships Windows 3.1 (16 bit) with a graphical user interface. One million copies of the new and upgrade versions are sold almost immediately.
1995	Microsoft releases Windows 95 (32 bit) with heavy advance advertising.
1995	IBM announces PC DOS 7, with data compression.
1996	Microsoft releases the Windows NT 4.0 OS designed for the business user.
1998	Microsoft releases the Windows 98 OS designed for the consumer.
2000	Microsoft releases the Windows 2000 OS designed for the business user.
2001	Microsoft releases the Windows XP OS designed for both the business user and the consumer.

Windows 98 and Windows Me

Both Windows 98 and Me are designed for the home user or casual user. Although these operating systems will work in a networked environment (LAN), they are not as secure as the NT and Windows 2000 operating systems.

Both OS's have plug-and-play features. Plug-and-play equipment, which installs and self-configures immediately, can be purchased. Equipment such as modems, printers, cartridge drives are all examples of plug-and-play equipment. This equipment is installed with little interaction from the user according to the manufacturer's own design and system capabilities.

If you intend to use or are using Windows 98, be sure to have the second edition of the software. The second edition of the Win 98 OS has the latest available fixes and software drivers supplied by Microsoft.

Windows 98 has a multiple user capability. Users have their own settings and password. The problem is that anyone can log in by choosing a new username and password. This is not a very secure solution for the business environment.

Windows Millennium Edition (Me) looks exactly the same as Windows 98. There's nothing visually different. However, some features are worth mentioning, for example, a system restore feature that lets you return to previous settings whenever you are having problems. Another new program, called the Windows movie maker, allows you to import, edit, and produce your own home movies on the PC. Windows Media Player is also included in this package. I review Windows Media Player in the next section of this chapter.

Windows Me contains system file protection (SFP), which prevents you or a virus program from actually deleting Windows core system files. Another feature is the system restore tool, which allows you to go back to a previous configuration without serious problems. The Windows Me start button displays programs that you've recently used unless you click on the down arrows (Figure 8.1).

Microsoft has also incorporated an automated OS software update feature in Windows Me. The OS checks for critical updates and gives you an option of downloading and installing them using the Internet.

Both Windows 98 and Me are designed for the consumer and not for the business user. That doesn't say that Windows 98 and Me won't work in a networked environment. They actually provide some networking and security features, however limited. For example, password protection is easy to bypass; file and print sharing must be turned off. The latest Windows releases should be employed to plug potential security holes.

Windows NT and Windows 2000

Both Windows NT and 2000 are designed for the business user and not for the consumer. All current Windows OS's have a similar look and feel, but there are major differences between each (Figure 8.2).

Windows NT is the most popular OS for business use, and it is the most stable (fewest errors) of all the Windows OS's. NT is not a plug-and-play OS. Therefore, it may sometimes be difficult to install new components or devices with the NT OS. Also be sure that your NT OS has the latest service pack loaded; it is provided by Microsoft as a free download.

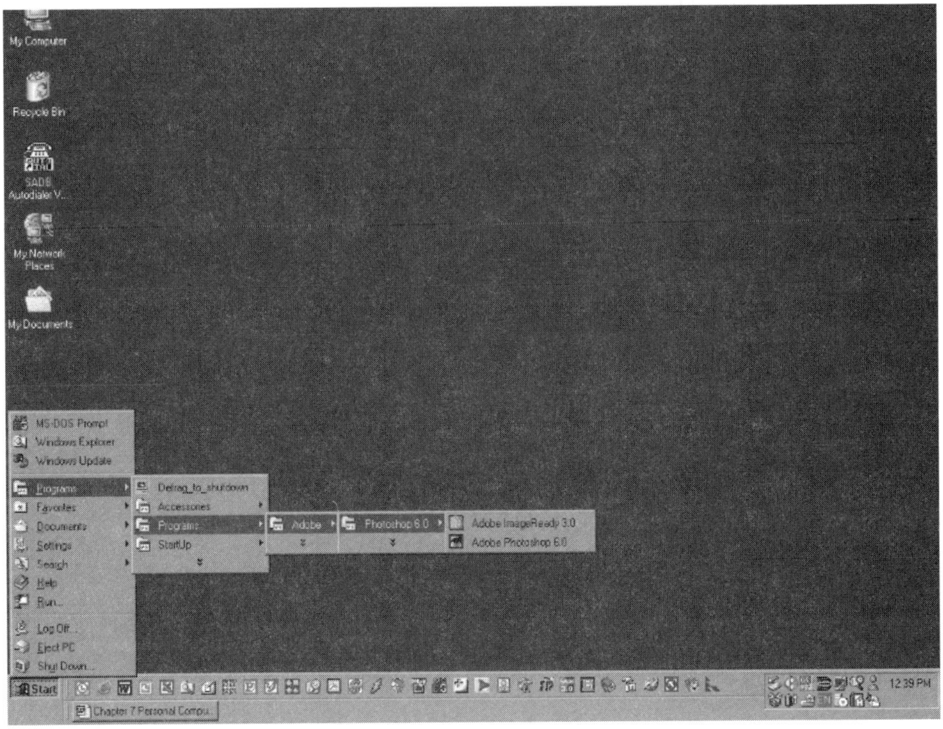

Figure 8.1
Windows Me GUI.

Windows 2000 has plug-and-play support and a consumer Windows look and feel. The OS is stable and has additional security features like an encryption file system (EFS) that enables you to encrypt individual documents locally on the hard drive.

Both Windows NT and 2000 provide multiple users with varying rights and privileges. For example, only users with administrative rights can install programs on

Figure 8.2
Windows NT and Windows 2000 logos.

Figure 8.3
Windows XP logo.

their PC. Users without administrative rights can only run programs and not install them. A LAN or PC administrator assigns the rights to each user, providing the correct level of access. Because most users do not need administrative rights to perform their daily business function, a higher level of security is obtained.

If you aren't sure which operating system to install, you should closely examine Windows 2000 or Windows XP for networked, shared business use.

Windows XP

The newest Windows OS is available in two versions: Windows XP Home Edition and Windows XP Professional (Figure 8.3). The general look is different from its predecessors, however; the OS essentially uses Windows 2000 code, which offers better reliability than does Windows 98 or Me. This OS also uses plug-and-play technology and provides far broader hardware and software support than does Windows 2000.

In Windows XP, the desktop has fewer icons, the Start menu shows only the applications you use most frequently, and commonly accessed dialog boxes are streamlined. The overall effect is a less cluttered OS. You can also change the desktop appearance to look like the old Windows. This is referred to as the Windows "Classic" look.

Many of the features in Windows XP, including plug-and-play support, are familiar to users running Windows 98 and Me. Microsoft has grouped software together, making existing programs easier to find and use by providing a task-oriented approach. Some of the new features in Windows XP, aside from the look and feel, include Internet Explorer 6.0 and Windows Media Player 8.

Microsoft has added some new OS setup requirements for Windows XP. You must install and activate the product within 30 days of purchase. The OS setup procedure checks the IDs of 10 different hardware components, allowing you to install the software on one PC only. If a particular component crashes, you may need to reinstall or reactivate your OS while online.

Apple

Apple Computer, Inc., has a long history in the world of personal computing. Their latest release, OS X, promises to be more stable than previous Mac OS's. Fast display of text and graphics and better memory management means fewer problems. The Mac operating system uses UNIX commands with openGL standards so that third-party developers can add applications more easily.

Apple computers are specifically designed for the Apple OS. Some users still find Apple's GUI easier to use than Windows' GUI.

Software created for the Apple OS is not compatible with software created for the Windows OS, and vice versa. Some emulator software packages exist that allow you to use Windows applications software with an Apple OS.

Timetable of Historic Apple OS Events

1975	Steve Jobs and Steve Wozniak work on the Apple 1 in Jobs' bedroom.
1979	Apple II and Lisa models are released.
1981	The Macintosh project begins.
1984	Apple Macintosh is released with a graphical user interface.
1985	Agreements between Microsoft and Apple allow Microsoft to use some of Apple's GUI technologies.
1994	Mac OS 7.5 is released.
1997	Mac OS 8.0 is released.
1998	The iMac, running at 266 MHz, is released along with OS 8.5.
1999	The Powerbook G3 laptop, running at 450 MHz, is released.
2000	Mac OS X is based on UNIX code. A PowerMac G4 with an Aqua cube design and running at 500 MHz is released.
2001	Mac OS X is officially released.

Linux

The Linux (pronounced LIH-nucks) operating system was created in 1991 by Linus Torvalds at the University of Helsinki, Finland. Linus continued working on his project until 1994, when version 1.0 was released to the general public. Linux is based on the UNIX operating system. UNIX is a well-established operating system spanning several decades.

Figure 8.4
Linux penguin.

Linux differs from Windows in its licensing arrangements. Linux OS licensing is developed under the General Public License (GPL), and its source code is available to anyone who requests it. Software developers, however, charge money for their version of Linux, keeping the original source code intact.

According to its creator, the Linux OS symbolizes functionality, availability, and flexibility while other popular OS's are too rigid and costly. The international symbol for Linux is the penguin (Figure 8.4).

Because Linux source code distribution is free, anyone can take a base OS model, add software, packaging, and documentation, and distribute it as his own. Examples are Red Hat Linux, Corel Linux, and Caldera Linux, with each organization including its own software.

Remember that Linux is neither owned nor controlled by anyone; however, Linus Torvalds owns the trademark and is still involved with source code development.

Before installing Linux as an OS, it is best to install one of the Windows versions first. WinLinux 2000 or PhatLinux will get you up and running in a Linux-like environment, and you need not remove Windows or do a full installation. Also note that commercially available software will enable you to run a different operating system by choosing one from a list at disk start-up time.

Utilities to Keep a Microsoft OS Running Efficiently

Microsoft Defrag or Disk Defragmenter is a software utility designed to keep the programs and data in order on the hard drive. Defrag places programs and data in a contiguous format instead of scattered throughout the drive. Once a disk is defragged, the read-write head will access the data faster and more efficiently on the hard disk drive.

Defrag first runs a scandisk utility, which checks the hard drive for errors and corrects any problems it finds. Once the disk scan is complete, the defrag program begins to run.

To run defrag in Windows 98 or Me, you do the following:

1. Click Start.
2. Click Run and type "defrag" into the drop-down program box.
3. Click OK.
4. Select the drive that you wish to have checked and click OK.

Defrag is analogous to a car tune-up. The car will run without a tune-up, but it must work harder and is less efficient. Therefore, why not use the Defrag version that comes with the Windows 98 or Me operating systems; it's free!

Bo Andersen's enhanced defrag program Defrag-to-Shutdown at *http://medlem.tripodnet.nu/mokkelmannen/defragtoshutdown.html* starts the defrag program with three options and with the option to automatically shut down the computer after defrag has finished. An option to disable the screensaver during the defrag process also exists (Figure 8.5).

Generally, unless the hard drive is seriously fragmented, you will not notice a significant difference in performance. However, to keep things running as smoothly as possible, it is good practice to run this program about once a week.

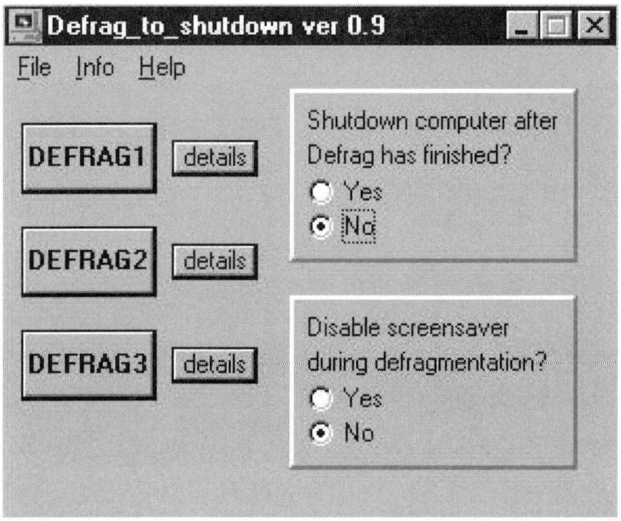

Figure 8.5
Defrag-to-shutdown. Courtesy of Bo Andersen (boimakra@hotmail.com).

Windows Screen Saver

Don't forget about the Windows screen saver password feature. This feature works with all Windows environments since Windows 95. Password protection is enabled after several minutes of inactivity. The exact number of minutes before a password is required can be varied by the user. This feature is useful in an office environment when you may be leaving your desk for a meeting, lunch, or project task.

APPLICATIONS SOFTWARE

An operating system without applications software is of little use for the business user or consumer. For this reason, OS's such as Windows come bundled with some applications software, but it is usually very limited. CD-writer applications software, data compression applications software (WinZip) or business applications software (MS Office) are examples of software needed for specific applications.

Many different types of applications software exist for both home and business use. Instead of reviewing some of the major applications software packages available today, I've chosen to review essential applications software required for business.

Office 97/2000/XP

The MS Office suite of software contains software specifically designed for the business user. With MS Word you create or modify documents; with MS Excel you create or modify spreadsheets; with MS Access you acquire database functionality; with MS Outlook you create or read email, and with MS PowerPoint you create or modify drawings and diagrams.

As you may well imagine, this suite of software tools is quite popular and surpassed its closest competition several years ago.

Each subsequent release of the MS Office suite from 97 to 2000 to XP employs new tools and new functionality. For example, the release of Office XP contains speech recognition. With speech recognition, you speak commands into a microphone, and the commands are recognized and processed by the PC. Other new features such as "Smart Tags" which recognizes data, document encryption, digital signatures, and document recovery features are used extensively in the latest version of MS Office.

TIMETABLE OF HISTORIC SPEECH RECOGNITION EVENTS

1968	2001: A Space Odyssey demonstrates the HAL-9000 computer with speech recognition.
1978	Texas Instruments introduces the Speak and Spell toy with limited speech recognition.
1993	IBM distributes the first speech recognition product for the PC; it's called the personal dictation system for OS/2.
1994	Dragon Systems releases Dragon Dictate for Windows.
1997	Dragon Systems ships Naturally Speaking as a general-purpose speech recognition product.
1997	IBM ships their Via Voice general-purpose speech recognition product.
2001	Microsoft releases MS Office XP containing speech recognition functionality.

The MS Office professional special edition suite provides additional software such as MS Front Page, with which you create web pages, and MS Publisher, with which you create documents that are suitable for publishing. Since not all businesses are the same, Microsoft has chosen to create professional, standard, and developer Office suites designed for users who do not require the entire package.

Windows Media Player

Although MS Windows Media Player is not designed for business, I'll cover the basics here. This software is Microsoft's attempt to capture the entertainment segment of the software market.

In a business sense, you may be required to view audio or video documentation files to learn about products or services. It is also possible that remote training will occur, using the transmission of streaming audio and video to save on training expenses. With the MS Media Player, it is possible to view local audio and video (from a hard drive) and streaming audio and video from the Internet.

MS Media Player software is based on DirectShow architecture designed as an entertainment product accessing multimedia content. This application comes bundled with the Windows Me OS, and it is also available as a free download for Windows 98 users.

Figure 8.6
Windows Media Player.

You can play many of the most popular multimedia formats, including Windows Media, RealAudio, and RealVideo 4.0 and below, QuickTime, MP3, AVI, WAV, and MPG, all from one easy-to-use application. Video features such as color, brightness, and hue are adjustable but not all formats are supported; for example, formats such as RealAudio and RealVideo are not supported.

Windows Media Player is like having a radio, movie player, CD player, and information viewer all in one. Use Windows Media Player to listen to streaming media radio stations, streaming video and audio from all over the world. Windows Media Player is available from the Microsoft download web site (Figure 8.6).

Microsoft is pushing Media Player as a home entertainment software package that has been met with great success.

Figure 8.7
NetMeeting.

NetMeeting

MS NetMeeting transmits and receives multimedia from another party by means of the Internet. This applications software comes bundled with the Windows 98 and Windows Me OS.

A connection with NetMeeting is a multimedia conference call. The images will be choppy when a dial-up modem is used. For cable modem and dedicated and xDSL transmission services, however, the video and audio is surprisingly good. You will need an Internet or web camera to capture the moving images. The cameras available for multimedia services are relatively inexpensive and serve their purpose.

NetMeeting has other features such as digital whiteboard, program and desktop sharing, and chat (Figure 8.7).

NetMeeting is also well suited for distance learning and multipoint data conferencing and is available from the Microsoft download web site.

DATA SAVING AND SECURITY

Generally, it is best to store all your data files in the My Documents folder that comes with Windows 98/Me/2000/XP. Most applications software options allow you to change the save setting where documents for a particular application are stored. These settings should be changed immediately to keep all data files together instead of in different file folders on the hard drive. If you are using Windows NT, you can create a "My Documents" folder yourself.

Saving and securing your data is one of the most important things you can do for your small business or personal files.

Sharing Files

If you are on a network, it is not a good idea to share data on your hard drive, especially if the data can be accessed without your knowledge or consent. All local drives should be made unsharable except in special cases. In a networked business environment, you can see your LAN or network administrator about this. Otherwise, check the properties setting for each local drive to be sure that the share has been turned off.

Zip Files

Zip files are used for distributing and storing files in a compressed format. Zip files allow for the grouping of several individual files into one, making transporting and copying files easier and faster. These programs are sometimes referred to as archive utilities.

A program called WinZip (*http://www.winzip.com*) makes it easy for Windows users to work with Zip files. WinZip features an intuitive point-and-click, drag-and-drop interface for viewing, running, extracting, adding, deleting, and testing files in a .zip format. The creation of these compressed .zip files is called "zipping"; the extraction and re-creation of these .zip files into their original state are called "unzipping." You can also add a password to zipped files so no one else can access them.

.Zip files will not compress .MP3 or .MPG or .AVI files since these multimedia files use their own compression algorithms. However, WinZip will let you group these multimedia formats together for distribution.

It is also possible to create files that unzip themselves. WinZip self-extractor personal edition is included with WinZip. Self-extracting files provide a good method

Figure 8.8
WinZip. Copyright 1991–2001 WinZip Computing, Inc. WinZip® is a registered trademark of WinZip Computing, Inc. WinZip is available from www.winzip.com. WinZip screen image reproduced with permission of WinZip Computing, Inc.

for sending multiple compressed files to others who do not own or know how to use .Zip software (Figure 8.8).

WinZip is one of the most important software add-ons for a personal computer. The registration and licensing costs are minimal and the software's benefits are substantial.

Backup

We've talked about backup several times when referring to hardware. Remember, a backup circuit or network is sometimes referred to as a redundant circuit or network. Well, backup is as important or more important when you look at data because you may not have the ability to recover data. You won't understand the importance of backup until you lose important files that you can never recover again. What can go wrong you say? Well, everything perhaps. Let me mention a few:

- Someone dropping or jolting a PC
- Static shock or electrical discharge (lightning)
- A virus attack (through email or newsgroups)
- A security breach (someone has your passwords)
- Water damage (burst pipe, spilled coffee, etc.)
- Fire or heat damage (extreme temperatures)

Actually, just a worn-out or malfunctioning hard drive is a mechanical device (covered in Chapter 7) that can put your business at risk. So what can you do? First of all, have a plan. This backup plan should include backup of important data files at least once a week. These important files should include any document or image that is difficult or impossible to re-create.

Remember, only the data needs to be backed up. You can always recover the programs themselves either by purchasing new install disks or by reinstalling the same programs again. However, if you feel that you want the OS and applications software backed up also, there are programs that will let you do this. Be sure that you have large enough media to handle this type of backup.

It is best to back up data to the My Documents folder as previously discussed.

Backup is basically a copy of the original data. You can use the following media to create a copy:

- Floppy drive—These disks are very small in capacity (1.44 MB), so they will probably not be large enough for your needs.
- Zip cartridge drive—100 Mbytes are adequate for backing up documents and text files but not large enough for backing up image and multimedia files.
- CD-R or CD-RW—650 Mbytes can be used to back up documents, text, and image files but probably are not large enough for multimedia files.
- Jaz cartridge drive—1-GB and 2-GB cartridges are adequate for backing up documents, text, images, and multimedia files.
- Another hard drive—External hard drives are available for backup using a FireWire interface that can hold up to 80 GB of data!
- A local network drive—If you have access to a LAN, you may want to back up your data to the server.
- An Internet network drive—Some Internet sites offer backup services for a fee. Check it out before going this route because of security concerns.

Oh, yes; before I forget, we should discuss a long-term backup. For example, a backup copy was not created correctly or became scratched or damaged. Let's say that you created a backup CD-RW and placed it in your locked desk drawer where it would be safe. Then an unfortunate catastrophic event such as a fire or a water leak damaged your PC and your backup copy. Everything is gone! You followed all of the rules and you wound up with nothing.

To compensate for these types of events and other catastrophic events, you should have an off-site backup plan in place. If up-to-date financial data is critical to your business, you should consider a company such as Comdisco to handle your backup needs. Other off-site backup plans could be as simple as taking home a disk from the office each night with the data that was processed during the day.

The World Trade Center disaster in 2001 emphasized the need for better security and data recovery. Many World Trade Center businesses, both large and small, were suddenly faced with the loss of data and the overwhelming loss of life. Many of these businesses are still struggling to recover both emotionally and financially while others were more fortunate and better prepared and could operate online again within a few days at remote locations.

Microsoft Backup is a software utility program included with the Windows 98 and Windows Me OS's. The program is simple and contains support for all of the media devices listed above (Figure 8.9).

Filtering Software

The Constitution of the United States gives us the right to freedom of speech. Some say this right is abused and misused. Regardless of where you stand on the issues of freedom of speech and freedom of the press, the Internet and email are here to stay. So we must decide what is viewable or objectionable for ourselves and our families.

Filtering software attempts to provide some control over the content that can be viewed. Web sites and email, including attachments, are scrutinized by the filtering software when transmitting or receiving data. A filtering software administrator (usually a parent) can decide on what information is objectionable or not objectionable.

These software packages basically check words and web site names against a master list of items not to view. If there is a match, the word is replaced or X'ed out. Similarly, if a web site address matches the list, the contents of the web site will not be displayed. Many of these lists are customizable and can be made more or less strict according to how the filtering software administrator wants to set the lists: very strict, not too strict, not very strict, etc.

New web sites are created each day, so it would be impossible for the filtering software administrator to update the list and make changes. The software vendor must become aware of the new web sites either by email notification from a user or another vendor. If the filtering software vendor deems the web sites objectionable, the site is added to their list. The next time a user accesses the Internet, the list is updated to be in synchronization with the filtering vendor's list.

Filters work also with data transmission. For example, filtering out personal information such as a child's name, address, and phone number will ensure that this information is not divulged to strangers.

Filters are generally not used in businesses because of the additional resources (both employee and financial) required. Employees are expected to follow company policy and access work-related email and web sites only.

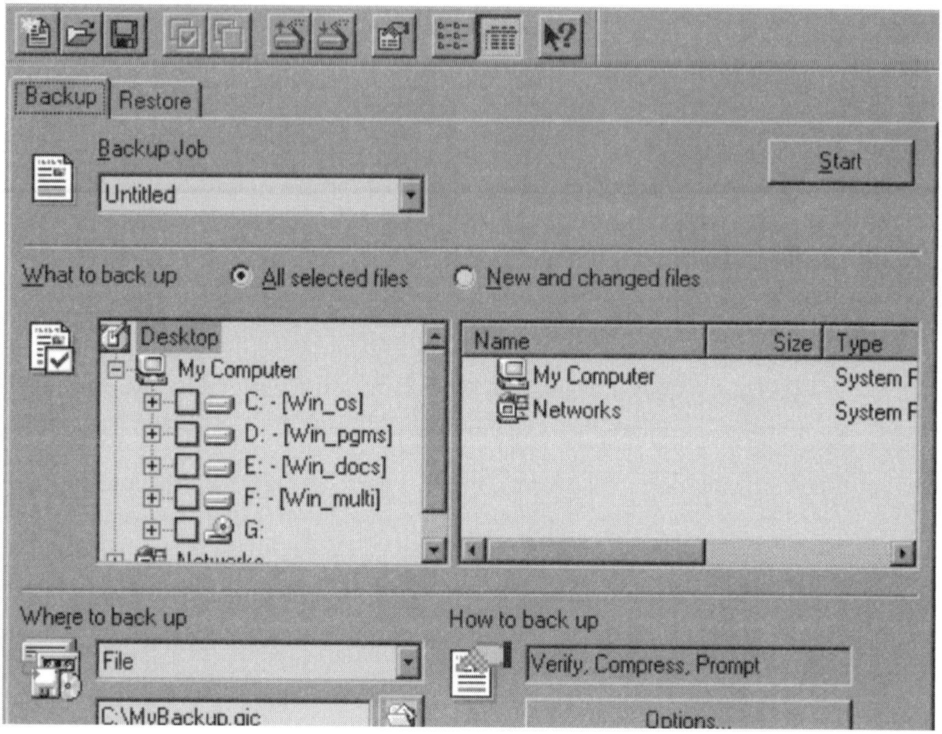

Figure 8.9
MS Backup.

Today's filtering software is not perfect. Legitimate web sites may be filtered because of a word or words used on a site that are seemingly not in good taste. More technology research and government legislation is needed to provide better protection for families using email and the Internet.

Some of the more popular filtering software in use today can be obtained from the following sources:

CyberPatrol: *http://www.surfcontrol.com*

NetNanny: *http://www.netnanny.com*

Cybersitter: *http://www.cybersitter.com*

Passwords

You may wonder what is so important about a password. A password provides access to systems and services. Anyone have Bill Gates' network or web site passwords? See what I mean?

A password is a sequence of characters and/or numbers that identifies who you are. Passwords are used to log in to a network, to establish a dial-up connection, to gain access to a web site, to secure important data, etc. However, if passwords are too easily discovered, someone can gain access to your private data that should not be accessible by anyone but you.

At one time, my passwords were very simple. "Beth01" (my wife) was a good one. "Penny01" (my dog) was another one. These are very simple and easy to crack. Why not try a password with a pass-phrase such as "The Beatles broke up in 1970". The pass-phrase would then be TBBUi1970 (No, I am not using this password on my PC).

Here are some recommendations:

Do's

- Use nine or more characters for passwords (more characters = better security).
- Use punctuation marks, symbols, and numbers for passwords.
- Use both lowercase and capital letters for passwords.
- Use pass-phrases mixed with special characters for passwords.
- Change your passwords frequently.

Don'ts

- Don't use dictionary names for passwords.
- Don't use names, birthdates, or places for passwords.
- Don't write your passwords on paper.
- Don't reveal your passwords to strangers or friends.
- Don't allow the Windows OS to save your password.
- Don't use the same password for several different applications.

There are databases that are specifically designed for secure password storage. One main password is used to access all passwords needed for web sites, databases, and personal information. A company called Counterpane Internet Security, Inc., has

Figure 8.10
Password Safe. By permission of Counterpane Internet Security, Inc.

a program, called Password Safe, which is downloadable from the web (*http://www.counterpane.com/passsafe.html*) (Figure 8.10).

A program such as Password Safe is a simple, cost-effective, and secure method of saving passwords. One word of common sense: Close this program while your PC is unattended.

If you don't care to use a vendor-supplied program for passwords, you can save a list of user names and passwords in a .zip file containing a master password.

You should be aware that several password-cracking programs are available today. Some programs simply try different combinations of letters, characters, and numbers until the desired result is obtained. This is why the length of a password is important. The longer the password, the longer it takes to crack. Other password-cracking programs include a dictionary-checking password program combined with a name-checking password program.

Quicken

Quicken 2002 Home and Business software lets you manage your home and business software in the same place, saving you time and effort. This software has the standard features of Quicken Deluxe, plus the key tools you need to manage a small business. You can do the following tasks with Quicken:

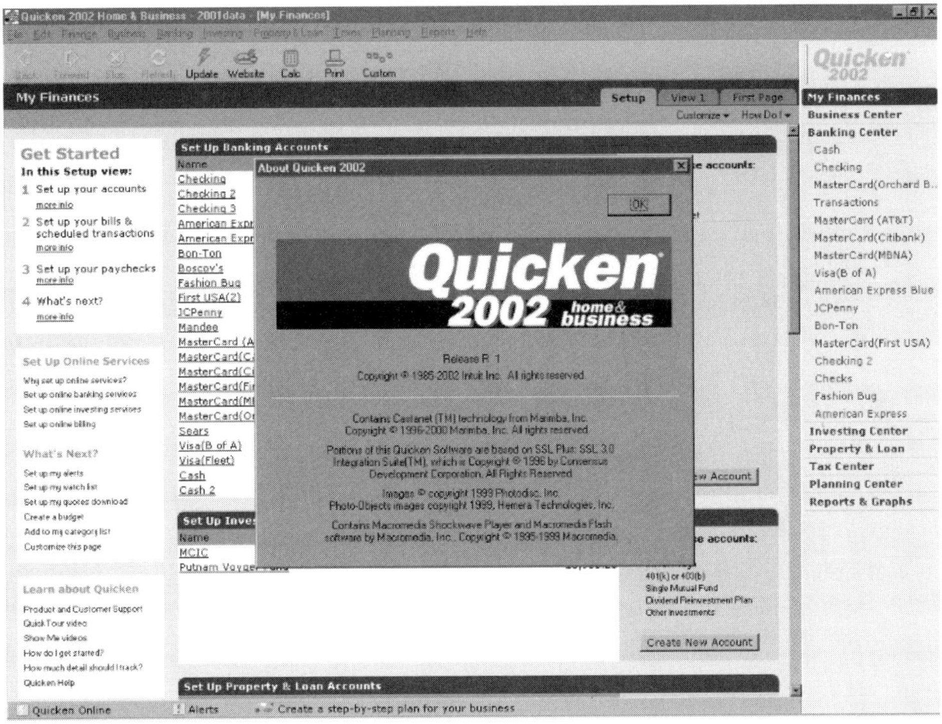

Figure 8.11
Quicken 2002 Home and Business. By permission of Intuit.

- Create and print professional invoices with your logo or on your letterhead and email them. Create estimates the same way.
- Automatically track expenses and assign categories like IRS Schedule C.
- Handle payroll requirements with an integrated payroll service.
- Easily manage accounts payable and receivable on a cash or accrual basis and see at a glance where your business stands.

Quicken software works with over 1,000 financial institutions, making it easy to bank online, pay bills, transfer funds, and download transactions from your bank credit card or investment accounts. The list of financial institutions is continuously updated by Intuit Corporation (Figure 8.11).

This financial software protects the privacy and security of your data. Quicken Online financial features take advantage of state-of-the-art encryption and authentication technologies, including Secure Sockets Layer (SSL), digital certificates, and RSA encryption to protect your information.

EMAIL AND PRIVACY

Electronic mail, or email, has become extremely popular with businesses and consumers alike. Email is sent through TCP/IP over the wide area network to both regional and international destinations in a matter of minutes.

Email is a store-and-forward service. A message is sent to an email server where it is stored for later retrieval. The message itself is a documented record of what has transpired. Voice conversations over the telephone cannot be retrieved unless recorded. Email messages can always be retrieved for further investigation.

An email program enables you to create a message, using standard text or HyperText Markup Language (HTML). Using HTML, you can incorporate images and enhanced text into your document before sending it.

Once the message is complete, it is automatically transferred to the outbox while the email program looks to send the message to its destination server. The message stays in the outbox until it is sent, and then the program automatically transfers it to the sent box.

Incoming messages are retrieved from an email server into your inbox. You can set the message retrieval process to be automatic or to occur when a button is pressed.

Attachments containing documents, images, etc., can be sent with email. Some email servers have a limit to the size of attachments, so check with your ISP before sending.

Not all email is wanted or necessary. Some email can be hoaxes. Some can contain viruses. There are ways to isolate and eliminate these bogus email messages, as we shall soon see.

Outlook

MS Outlook is part of the MS Office suite of business productivity tools. Outlook is a scheduler, contact list, and email organizer. The beauty of Outlook email applications software is that it is fully integrated with the other MS Office products, including MS Word, MS Excel, and MS Access.

With Outlook, you can send messages from your calendar and perform searches relating to a specific event. You can use the features of MS Word, including spell checker and grammar checker, to name a few (Figure 8.12).

Outlook XP contains additional features such as the email box cleanup tool and the email archiving tool. In addition to these new features, Outlook XP has some virus protection features. Attachments containing extensions of .BAT, .EXE, and .HLP can-

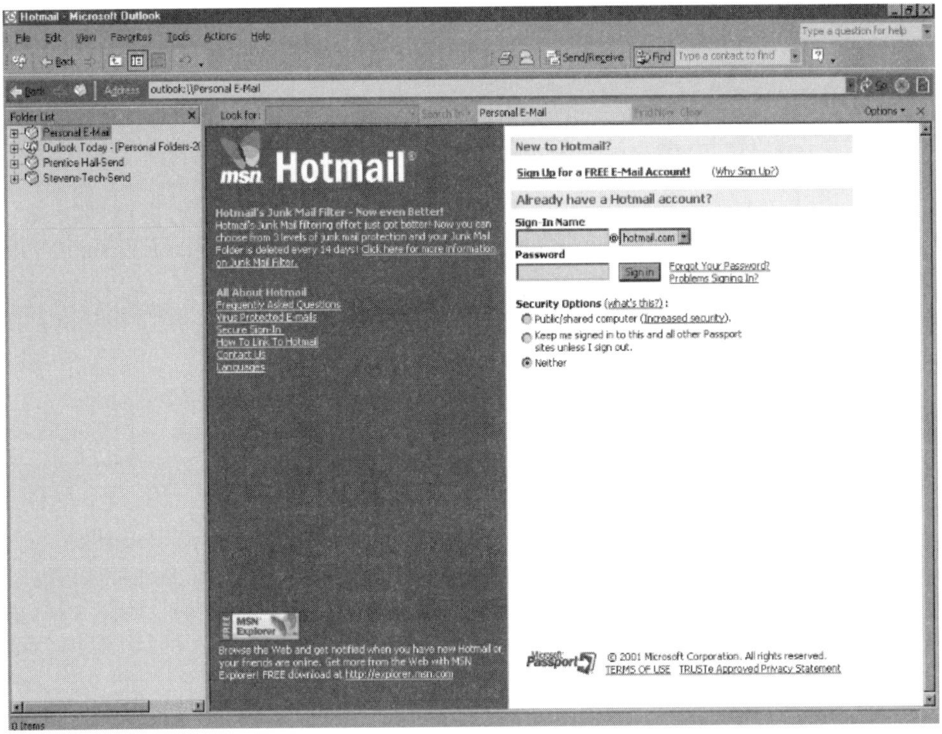

Figure 8.12
MS Outlook.

not be opened. If someone needs to send you a file containing these extensions, they must send you a .zip file.

MS Outlook is today's standard applications software for sending and receiving email.

Visual Basic for Applications

Visual Basic for Applications (VBA) is the programming language used for all MS Office applications. Users can create macros, which are stored as VBA programs that can be executed and edited just like standard Visual Basic programs. By creating VBA macros, users can extend the functionality of MS Office applications within their working environment.

The functionality of VBA has enabled users to customize workflow applications to meet organizational needs. Along with the increased flexibility that VBA provides, there is some increased vulnerability because VBA is frequently used to create macro viruses. We'll review more about viruses in the next section.

Pseudonymous Email

Pseudonymous email is another name for an anonymous email account. This type of email provider basically strips off the sender's information before letting the packet continue on its way. If the intended receivers need to reply, they must go back to the pseudonymous email provider, who can send the message back to the original sender.

The problem is that these service providers know who you are and how to contact you in case of an illegal activity. There are several examples where law enforcement has ordered a pseudonymous email provider to provide account information revealing the user's identity.

Spam Email

Spam email is also known as junk email. The message is sent in bulk by specialized marketing companies to a large group of email recipients. This practice wastes resources and IP bandwidth. Some ISPs screen for spam by reviewing the email messages first and then deleting a message if it is spam email.

OK, so you may ask: How do I know this is a spam email? Well, for one thing, the return address isn't usually valid. This type of unsolicited junk email usually contains advertised products. Chain letters, stock offerings, get-rich-quick schemes, health remedies, prize promotions, etc., are some examples of spam email.

Spam email is here to stay because of low production costs and the fact that these junk email messages reach thousands of people in an extremely short time.

Many email software packages offer an automated method of filing spam email in a junk email box so it is not received with your regular email.

Digital IDs and Encryption

A digital signature is used with email as a piece of the message containing encrypted data. The encrypted data contains public key information with the ability to ascertain that the message is true and authentic. Public key information decrypts two keys. One key is used for public view and the other for private view. Either key can encrypt a message, but only the private key can decrypt it.

The private key must be distributed to all intended recipients of the email so that the messages can be decrypted. Note: For improved security, the larger the key, the better the encryption. A 64-bit key is very secure, and 128-bit keys are becoming increasingly popular.

A private key can be used to authenticate a person's identity. For example, if I send you an encrypted email, you know that the message is coming from me and that no one else is pretending to be me. Well, what if I set up public and private keys and call myself Mr. George Bush? First of all, the best method of determining whether I'm Mr. George Bush is to use a Certification Authority (CA). A CA is a trusted organization that authorizes and verifies public keys so I can't claim to be someone who I'm not. The CA keeps a copy of my keys, so you don't have to send it every time you need to send an encrypted message.

Digital IDs are the subject of privacy issues. Does the government have right to view data that has been encrypted? Should countries supporting terrorism and illegal activities be allowed to use encryption techniques to conduct their business? International cryptography laws have been created to keep threats against U.S. interests at a minimum.

Mailing Lists, Newsgroups, and Chat Rooms

Mailing lists are used to distribute your email information to a wide audience—sometimes hundreds or thousands of recipients who may share (or not share) the same interests that you do. Your email message is therefore sent in a bulk format, and the message needs to be sent only once. Your message might be referred to as a posted message.

If you are also listed as a mailing list recipient, you may receive several messages a day from other members of the same mailing list. Many of these email messages may be totally useless or even hostile.

Some mailing lists are moderated and read by a person before they are posted. The moderator's job is to approve the posted email message before it is distributed. A moderator might also check to see which emailer is creating the most problems and then permanently remove that account from the mailing list.

Moderators are frequently the target of email attacks themselves, and the position is usually voluntary.

A newsgroup or bulletin board is similar to a mailing list but does not use email. You can post messages to a bulletin board where all who wish to read them can do so.

Similarly, you can read any posted messages you desire without receiving unwanted or undesirable email each day. The best part about a newsgroup is that you do not have to belong to the group to view messages. Again, newsgroups can be monitored just like a mailing list.

Chat rooms allow users to meet and exchange information with other users in an open environment. The problem is that a person's real identity is masked. In reality, you may be talking to a stalker, criminal, deviant, or perhaps a law enforcement official.

You should realize by now that a false name or fictitious identity can be easily traced from an IP address that the PC generates to make the chat room connection. Therefore, all of the services we mentioned—mailing lists, newsgroups, and chat rooms—are not secure at all. It is just an illusion that these services are private, secretive, and confidential.

Some companies prefer to have these services blocked because they serve little purpose in the business world. If used properly in an open forum, however, these services can be fun, interesting, and entertaining.

BROWSERS AND PRIVACY

Browsers are designed to display both text and graphics using HyperText Markup Language (HTML). Presently, two browsers are available: Internet Explorer (IE) and Netscape. Our focus is with IE (Figure 8.13) because it is bundled with the Windows OS and because the functionality of both browsers is similar.

TIMETABLE OF HISTORIC BROWSER EVENTS

1993	Mosiac 1.0 is released.
1994	Netscape 1.0 is released.
1995	Mosiac 2.0 and Internet Explorer 1.0 and 2.0 are released.
1996	Netscape 2.0 and Internet Explorer 3.0 are released.
1997	Mosaic 3.0 is released (Final).
1997	Internet Explorer 4.0 is released.
1998	Netscape 4.5 is released.
1999	Internet Explorer 5.0 is released.
2000	Netscape 6.0 is released. It integrates browsing, email, and instant messaging and includes password manager and cookie manager software.
2001	Internet Explorer 6.0 is released. It provides privacy management tools and an error reporting tool.

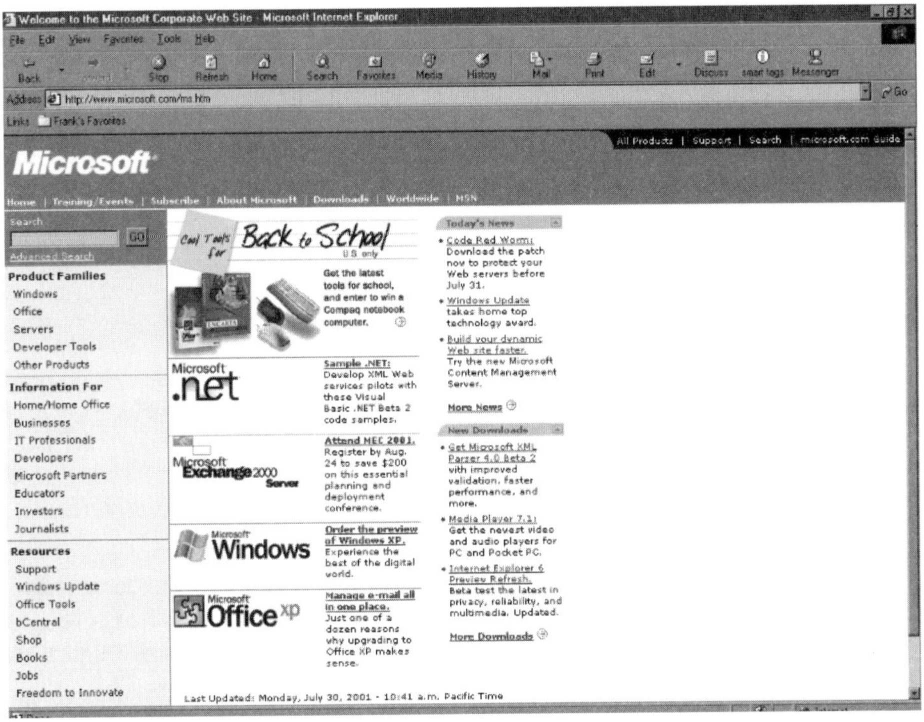

Figure 8.13
Internet Explorer.

We already know that the browser and the wide area network (WAN) have changed computing for good. No longer do we need to request information from an equipment supplier by regular mail. Today, we can simply go to its web site and download information immediately.

How much security and privacy do you have when surfing the Web? The answer is "not much." Information regarding web sites visited is available on your PC, to your ISP, and when at the workplace, to your employer. This information may not be available to the next-door neighbor but may be of interest to law enforcement or your employer. You should always keep in mind that a record of your conversation or viewed web site is stored somewhere for someone else to view.

Personal Computer Cleanup Time

We cannot control what information we leave with our ISP or our employer, but we can control what information we leave behind on our PC. Much of this information is useless or unnecessary for the average user. Looking at MS IE 5.0 and up as an example, we should pay attention to the following five areas:

- History—Contains links to pages you've visited. Go to the Internet properties -> general tab -> history section. The default setting is 20 days. Set this value to 5 days or less.
- Auto Complete—Stores address locations of sites you've visited. Go to the Internet Properties -> Content tab. Click on the web address check box to uncheck it.
- Temporary Internet Files—These files are stored in the Temporary Internet Files folder under Windows. This temporary information, which includes both images and text, is used when you return to a web site you've already visited. When this happens, the information is retrieved locally from your hard drive instead of from the Internet. You can clear this information by going to Internet Properties -> general tab -> temporary Internet files section. Click on the delete button to remove.
- Cookies—These files are stored in the Cookies folder under Windows. See the "Cookies" section below for details on what a cookie is. You can simply highlight all data in the cookies folder under Windows and delete.
- Favorites—Some web sites automatically add their site to your favorites menu. To delete these, click on the Favorites menu and right-click on the link you want to remove. Click on delete to remove the link.

Personal Information Access

It is possible for a visited web site to request personal information about you without your knowledge or consent. The IE 5.0 and above browser has a feature providing information from your address book to any web site that wants the information. I find this particularly annoying because, again, this is done without your knowledge or consent.

It is possible, however, to disable this feature as follows:

From the Start, Settings, Control Panel:

1. Go to Internet Options.
2. Go to Content Tab.
3. Go to My Profile button.
4. Enter a fictitious name.

Cookies

A cookie is a snippet of data that is stored on your hard drive when you use a browser to surf the Web. This data can include your name and preferred settings from your previous visit. When you revisit the web site, the browser will search for any cookies on your hard drive. If the browser finds a cookie, it customizes the web site in accordance with the cookie information. The site will already know who you are through the use of the cookie.

Cookies are relatively harmless because data is held in the cookie and not in executable program code. However, always keep in mind the following privacy issues:

- Cookies can follow the type of web sites you've visited, allowing companies to target you with a certain profile.
- Cookies can contain specific information that you may not want others to use. For example, if you share your computer with a coworker or family member, information about you may be displayed on the browser.
- Cookies leave a record of the sites that you have visited.

Many large commercial web sites called portals use banners for advertising. Examples of web portals are *http://www.go.com*, *http://www.msnbc.com*, *http://www.cnn.com*. The purpose of web site advertising is to generate revenue to help defray the costs of maintaining the site. Advertising companies create their banners, using cookies to obtain information. The cookie information can contain data such as how many times you've seen their ad and what types of web sites you are viewing. For example, if you like to go to audio equipment web sites, you may start seeing audio equipment ads the next time you see a banner at a portal's web site.

Some of the information obtained from cookies is sold for profit. The information is compiled into a detailed list of all web surfers searching for audio equipment (using our example). Again, all of this is done without your knowledge or consent.

To combat the privacy dilemma, anti-cookie programs are becoming available. They help you manage the amount of information disseminated. Also, the latest versions of Internet Explorer contain more cookie control than ever before.

E-Commerce

Shopping by browser is one of the most popular conveniences available. The Internet provides virtually limitless capabilities when it comes to accessing goods and services that you can pay for with a charge card. It is important to learn the risks involved in using your charge card on the Web.

Normally, when you are finished with your purchases, you click the send order button. Purchase information along with your name, address, and charge card information is packetized and sent through IP to its final destination. The problem is, anyone can view this data as it travels from router to router.

The only way to protect your information is to use encryption. Encryption provides the necessary security so that your information cannot be copied. Secure Sockets Layer (SSL) is the most common protocol used for secure web transactions. The data is unreadable as it is sent from browser to web server.

Another encryption protocol called Secure Electronic Transaction (SET) is used for credit card transactions. SET is actually less secure than SSL because it encrypts only the credit card portion of a transaction.

How safe is e-commerce? Well, credit card numbers can be stolen at retail stores and outlets, so you are at risk in both e-commerce and retail situations. Best advice is to keep credit limits low (less than $1,000) so if your number is stolen, the most you can lose is a few hundred dollars.

VIRUS ATTACKS

A virus is a program that performs a malicious task without your knowledge or consent. The virus then replicates itself and spreads to other computers, destroying data and further spreading from those PCs. You may have experienced a virus yourself or at least heard of someone, such as a friend or a coworker, having a virus.

Viruses can cause extensive data damage. They can also wreak havoc with an operating system, rendering it useless. Operating systems can always be reinstalled but data can be lost forever. There are many different ways to pick up a virus. A floppy with an infected file might be passed on to your PC. Email might have an attachment containing an infected file. A file server might be infected. And so on.

Virus hoaxes are also popular. Hoaxes sent through email warn of potential disasters, death, or injury. Hoaxes of potential viruses and/or timed data destruction cause panic and wasted time, but they rarely, if ever, occur.

Virus creators love to target popular products such as Microsoft, so beware!

To avoid viruses from entering on your PC you should take all of the following precautions:

- Install antivirus software.
- Keep antivirus software up to date.
- When using email, check the sender's address and email content before opening any attachments.
- Back up, back up, back up.

TIMETABLE OF THE MOST FAMOUS VIRUSES AND HOAXES

1969	Beatle Paul McCartney's death hoax was fueled by a disk jockey claiming that clues were in the band's music, cover art, and lyrics.
1988	"Jerusalem," which activates every Friday the 13th, is released. Both .exe and .com files are affected.
1988	"Stoned" is released as a boot sector virus. This virus displays a message upon system boot-up stating "This computer is stoned."
1989	Rock band Milli-Vanilli sold millions of records. However, the actual singers were different people. Rob and Fab admitted to the hoax and continued with a recording career but could not regain their success.
1989	Datacrime (called Columbus Day in the USA) is released. This virus triggers a low-level format of the hard drive, making the PC virtually useless. It was set to run on October 12th of each year.
1991	"Tequila," which is the first polymorphic virus, is released.
1994	The "Good Times" hoax predicts a virus that will erase a hard drive when a user opens an email.
1995	The "Concept" MS Word virus is distributed.
1996	The Church of Kurt Cobain was a hoax that claimed the Church would draw religious attention to Gen X'ers by using Nirvana's lyrics.
1998	"Chernobyl" activates itself each April 26, commemorating the disaster of the Chernobyl nuclear disaster on 4-26-86. When activated, the virus destroys the first Mbyte of data on a hard drive.
1999	"Melissa" executes a macro in a document attached to an email which forwards the document to 50 users in MS Outlook's contact list.
1999	"Love Bug" or "I Love You" virus performs much like "Melissa" but also deletes .MP3 and .JPG files.
2000	"Loveletter" (similar to the "Love Bug" virus) was allegedly written by a student in the Philippines. The attachment contains the words "I love you" and quickly spread, becoming the most destructive virus ever.
2000	"Anna Kournikova" virus operates similar to "Melissa" and "Love Bug." The virus is inserted by email containing an image of Anna Kournikova and spreads by sending copies of itself to the entire MS Outlook contact list.
2001	"Home Page" virus email redirects you to sexually explicit web pages.
2001	"Code Red" worm defaces MS web servers. Most likely released from China, this worm spreads rapidly, infecting many web sites.

Antivirus Software

Antivirus software examines files and directories on your PC for known viruses that may attack and destroy your PC. The software should also be capable of examining floppy disks (upon insertion) and RAM. In addition to the above, antivirus software should scan incoming email attachments to be sure that viruses do not enter through email software such as MS Outlook.

Antivirus software contains a database of known viruses. It must be periodically updated to ensure that all of the latest viruses will be captured. It is a known fact that several hundred viruses or variations of existing viruses are created each week. This database update, which should occur weekly or monthly, is critical to keeping your PC virus free and in good operating condition.

Virus-checking software uses PC system resources. But the disadvantages of slower OS response time are certainly minimal compared to the possibility in obtaining a virus that destroys your data.

Once the software detects a virus, a message informs you that a virus has been discovered in a program or data file. You basically have the option of deleting the infected data file entirely or cleaning the virus out of the data file and salvaging the good data. Your first option should be to clean the virus out. If this doesn't work, the program or data file containing the virus should be immediately deleted.

Viruses have several points of entry. Sometimes a network file server, an Internet gateway, or an email attachment is used to gain access to your PC. The email attachment is by far one of the greatest methods used to transmit a virus.

Keep alert and beware of all email attachments. If the attachment is from someone you don't know, don't open it. Also, if the attachment has .vbs, .hta, or .ocx extensions and you are using MS Outlook, don't open them. If you are unsure whether an attachment is virus free or not, perform the following steps:

1. Save the attachment to a floppy drive without opening it.
2. Delete the original email (with the attachment).
3. Write-protect the floppy disk and scan for viruses.
4. Open the document if no viruses are found.

Several types of antivirus software are available:

McAfee: *http://www.mcaffee.com*

Symantec: *http://www.symantec.com*

Command: *http://www.commandcom.com*

Macro Viruses

A macro virus contains a series of instructions usually written in VBA. These viruses can perform a task, usually with disastrous results. They spread by sending copies of themselves to others, using the victim's email contact address book.

Macro viruses attach themselves to documents containing data and can spread very quickly. These types of viruses might be written to be application specific, such as for MS Outlook and MS Word, but they are not usually operating system specific. In other words, the macro virus can be transferred between different operating systems. The macro virus is the most common virus you'll encounter.

If you are concerned about infection from macro viruses, it is a good idea to disable macros in MS Office products if you don't use them.

File Viruses

A file virus infects programs only and not data files or documents. Program files usually have an extension of .exe or .com in their filenames. The virus is spread when the program is run. Many viruses of this type are embedded in memory, affecting virtually every program installed on the PC. The only thing you can do upon encountering a file virus is try to recover your data and then reload the OS and applications software.

Boot Sector Viruses

The boot sector virus operates a section of the hard drive called the boot sector. The boot sector provides the CPU with information describing the hard drive's contents. All disks, whether floppies or hard drives, have a boot sector. A boot sector virus is particularly devastating because you can only recover by loading the antivirus software at boot time and attempt to eliminate the virus. Fortunately, most antivirus software checks for boot sector viruses at the time of boot-up.

Worm Viruses

A worm virus is designed to find as many PCs as possible on a local area network (LAN) and attempt to replicate itself onto the other PCs. Worm viruses use the network instead of email to propagate.

Other Types of Viruses

There are other types of viruses, some of which are more dangerous than others.

For example, a polymorphic virus is designed to fool antivirus software by changing its features to make itself undetectable. To find this type of virus, you must use sophisticated antivirus scanning techniques. This is a potentially dangerous virus because it is so difficult to detect.

Copycat viruses are slight variations of existing viruses that create what may be termed a new strain of an existing virus.

Another type of virus, the stealth virus, hides from the OS by residing in the PC's RAM. Still another virus, called the retro virus, attempts to disable the antivirus software needed to combat viruses in the first place.

Some viruses are combinations of all categories we've discussed. These multilevel viruses by design perform even more destruction, often with disastrous results.

SECURITY AND PRIVACY

Who do you trust? Is it a friend, neighbor, coworker, or family member? Depending on your situation, it may be all, one of the above, or somewhere in between. For example, would you leave your wallet with your personal information in a supermarket shopping cart and expect to find everything intact an hour later? Possibly not. Could a Good Samaritan find the wallet and return it to you? Sure. Great. However, if personal information has been copied, you may still be at risk.

In our above example, did you keep personal information such as mother's maiden name, social security number, or system passwords on a paper? If you did, you are at a greater risk than most for identity theft. Carelessness or outright theft can always happen. The question is, did you do your absolute best to secure data? A little bit of paranoia is not the worst thing to have. It may actually be beneficial in the long run.

In the data world, we must minimize the chances for theft and destruction of data by performing routine backups and using passwords. It is important to be knowledgeable regarding security and privacy issues. How do you view your personal privacy?

In the business world, employers have rights to your time and the data on your business PC, so the issue of privacy is a difficult one. How much data privacy is needed in the business environment?

TIMETABLE OF HISTORIC COMPUTER PRIVACY EVENTS

1999	Intel Corp. announces a new Pentium III processor that contains a processor serial number (PSN).
1999	Several consumer and privacy groups file a complaint against Intel Corp., alleging that the processor serial number would allow users to be tracked across the Internet.
1999	Intel Corp. bows to pressure and ships the Pentium III processor with the serial number turned off.
2000	Intel Corp. announces that future releases of the Pentium processor will not contain a PSN.

PCs and Data Theft

There are ways to circumvent password security, for example, during the PC's boot-up process. The prospective thief can bypass your OS security by tampering with the boot process and inserting another OS without security to view and copy data. The thief can cause even further damage by reformatting the hard drive and deleting data. You can foil these attempts by requiring a boot password.

Of course, it is always possible that critical internal components such as the CPU, memory, or hard drive can be removed. Again, most internal hardware is relatively inexpensive, so it is the data that must be safeguarded and proper backup and hard drive security (with a locked PC case) must be implemented.

Some manufacturers are specifying new products in an attempt to keep the problem of data theft under control. Fingerprint scanners, USB tokens, and eye (retina) scanners are all possible devices that can be used to counteract theft. Of course, the old PC lock and cable work well when these new security gadgets are either too expensive or unobtainable.

Keystroke Monitor Eavesdropping

One type of monitor used to invade privacy is the keystroke monitor. This monitor is a program that records every key pressed on a PC, including passwords. Someone with access to your PC can install one of these programs without your knowledge or consent. In many cases, antivirus software will detect and eradicate these programs, but the damage may already be done.

Packet Sniffing and Eavesdropping

A network monitor or traffic analyzer can be used as a packet sniffer. A packet sniffer is sometimes used for electronic eavesdropping. Analysis of packets would reveal the sender and intended receiver of a packet and how much information has been transferred. Generally, so many packets are transferred that it is difficult, if not impossible, to isolate the person or persons you are looking for.

For example, if an employer or an intruder can examine enough packets from the network, that person could find out who you are exchanging information with and for how long.

The best method of eliminating threats of privacy from a packet sniffer is to encrypt your email messages. However, even encrypted messages contain a sender's address and an intended receiver's address.

Capture-and-Modify Eavesdropping

Another security breach is the capture-and-modify attack. This type of eavesdropping involves a third party who intercepts your messages, modifies the data, and sends the altered messages on their way. It is also possible for the eavesdropper to capture and save your messages to be repeated over and over again to the intended party. Repeated messages can sometimes be used as a mailbomb attack.

Denial-of-Services

A denial-of-service attack floods an Internet server or network with repeated data or commands, thereby rendering the service unavailable. Tens of thousands of requests are made from one or several PCs to a single network or computer with an incorrect sender's address. Legitimate users needing to gain access to a network or server cannot do so.

Denial-of-service attacks on PCs can also be made through an email attachment that runs a program. This program is created to tie up the CPU with unnecessary computations, making the PC virtually useless.

One other denial-of-service attack is the mailbomb attack. This attack uses email to overwhelm a server or an email address with useless large email files.

CONCLUSIONS

We've taken a look at some basic, business-related software in this chapter. Different operating system software was reviewed along with some basic business applications software.

The other focus of this chapter was to learn about security and privacy. Without the proper tools and knowledge, you place your business or personal data at high risk. You should determine your level of risk and take appropriate action in the form of increased backups.

In the next chapter we again look at hardware devices. These devices may be necessary or even critical for operating in a small-business environment.

9 Local Equipment and Networking

In the last two chapters we concentrated on hardware and software for the Personal Computer. The PC is such an integral part of today's business and consumer environment that two chapters were dedicated solely to this technology.

This chapter begins to bring some of our knowledge about networks and equipment together to emphasize products and features that can help enhance your business. The chapter is divided into five sections.

First, we discuss local equipment requirements. Local equipment for a small business includes voice call management. Much of this equipment is customizable. If your business requires a help desk, you may be looking for more sophisticated local switching equipment. A smaller business, however, may be looking for local equipment that will perform basic functions.

Section two covers wireless devices. Pagers and cordless telephones have become so commonplace it would be remiss not to review them.

Section three focuses on imaging devices. We've already looked at video terminals and television imaging equipment. Here, we take a look at some imaging devices, such as scanners, digital cameras, and video projectors, that are gaining in popularity.

Section four takes a look at security for small businesses. Alarms, keys, and proximity switches are used to secure office environments.

Last, in section five, we revisit the PC and take a look at a typical networking setup.

More gadgets, gizmos, and toys…

TRADITIONAL VOICE EQUIPMENT DEVICES...........

This section covers the multitude of devices and attachments that are available for voice services. The equipment covered here ranges from simple caller-ID to more complex devices such as voice servers and computer telephony integration (CTI).

Telephone equipment vendors are all competing for this very lucrative market that can save tremendous sums of money for both small and large businesses while making a call center more efficient.

If you are not interested in purchasing and maintaining the equipment required for these devices and attachments, your local telephone company will be more than happy to provide you with advanced services for an additional fee.

Private Branch Exchange

A private branch exchange (PBX) is a private local switch usually purchased by a customer to handle voice communications within a building or complex. Generally, a PBX is designed for a business with 100 or more users. At a basic level, a PBX facilitates the following:

- Internal telecommunications between coworkers and individuals
- Telecommunications from individuals inside a building or campus to the PSTN
- Telecommunications from the PSTN to individuals within a building or campus

PBX switching eliminates the need for an individual local loop for each telephone. Internal calls are handled within the building or campus and are never connected to the PSTN. Outside calls or calls to the PSTN are handled through trunk lines connected between the PBX and the central office (Figure 9.1).

Several advantages and disadvantages are associated with a PBX:

Advantages

- Internal network—Individual local loops are not required.
- User maintenance—Changes can be made quickly without local telephone company involvement.

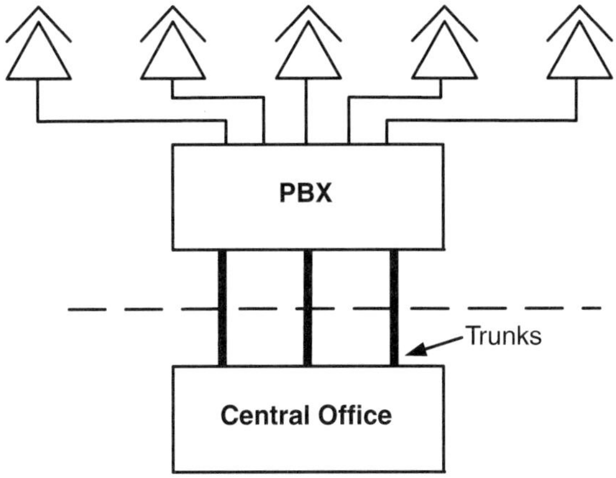

Figure 9.1
Private branch exchange.

Disadvantages

- User maintenance—A PBX administrator is needed.
- Proprietary equipment—Parts, telephones, and memory must be purchased from the same PBX vendor.

There is no such thing as a standard PBX. Each vendor manufactures its own proprietary line of PBXs. You must buy telephones that are supported by the PBX manufacturer.

PBX features vary from manufacturer to manufacturer. In addition to providing dial tone, a PBX may have features such as voice mail, call forwarding, speed dial, message waiting lamp, call waiting, call conferencing, etc.

Key Telephone System

A key telephone system is similar to a PBX. Generally, a key telephone system is a smaller version of a PBX, usually serving a small business with fewer than 100 employees. Many new homes or homes with advanced telephone wiring have these systems installed. With a key telephone system, all features such as voice mail, paging, etc., are centralized along with the wiring.

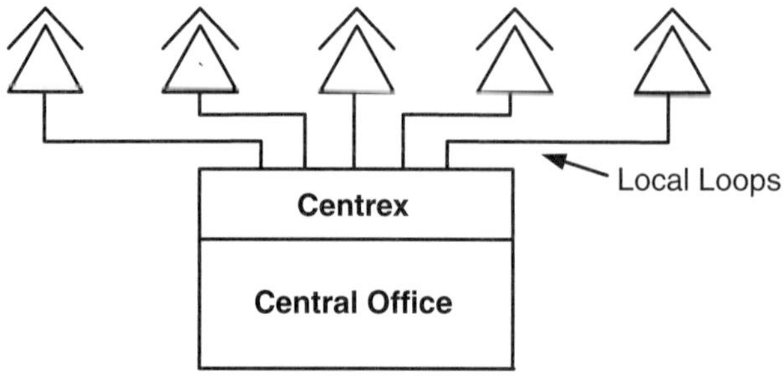

Figure 9.2
Centrex.

Check for key telephone features and pricing and compare them with PBX features and pricing. Again, there are no manufacturing standards for key telephone systems.

Centrex

A Centrex, or central exchange, is a local telephone company switch that handles voice communications within a building or complex. At a basic level, a Centrex switch accomplishes the same goals as the PBX, which are the following:

- Internal telecommunications between coworkers and individuals
- Telecommunications from individuals inside a building or campus to the PSTN
- Telecommunications from the PSTN to individuals within a building or campus

Centrex switching requires an individual local loop for each telephone. Internal calls must first travel to the central office Centrex and then back to the building or campus. Outside calls or calls to the PSTN are handled normally (Figure 9.2).

Several advantages and disadvantages are associated with a Centrex switch:

Advantages

- Local telephone company maintenance—No user maintenance required
- Automatic upgrades—Software and hardware enhancements automatically provided
- Monthly bill—Monthly charges

Disadvantages

- Individual local loops—Local loop required for each telephone device
- Change management—Changes processed by the local telephone company (may not be made immediately).

In a Centrex situation, the LEC may install a T-1 dedicated line instead of local loops. Remember, a T-1 line carries 24 simultaneous digitized voice channels, uncompressed.

Automated Call Distribution

Automated call distribution (ACD) systems are used in the business world to streamline call center operations. An ACD system efficiently routes telephone calls, based on customer DTMF keypad entry, to an agent. The ACD also collects call history data and can produce management reports to enhance and streamline a call center process.

At one time, ACDs were developed for large business organizations and corporations and cost thousands of dollars. Today, ACDs are small (about the size of a desktop PC) and are relatively inexpensive (about the cost of a large PC server). Some features of ACDs are as follows:

- Efficient agent staffing—Based upon call volume and ACD management reports
- Increased employee satisfaction—Relatively even distribution of the workload
- Increased customer satisfaction—Providing customer service at its highest level
- Reduced telephone expenses—Decreased time needed on the telephone because of automatic call routing
- Automated call volume reporting—Providing management with day-to-day reports as well as future expectations.

When a call comes into an ACD, the ACD can perform one of several routing methods. For example, the agent with the longest idle time may receive the next call. It is also possible to program the next call by using a balanced call count where the incoming call is sent to the station with the fewest calls.

Voice Response Units

As consumers, we encounter a voice response unit (VRU) almost daily. Banks, service-oriented companies, retail outlets, all use VRUs. VRUs use DTMF tones from the caller's telephone keypad to route incoming calls internally. A digitally created computer voice prompts the caller to respond by pressing the appropriate number on the keypad.

When a customer calls, a VRU can prompt for an account number and provide an automated account balance.

VRUs can be integrated into PBX or ACD terminals.

In addition to making better use of an employee's time, VRUs provide 24 hours of customer service. This provides more time for growing a core business.

Computer Telephony Integration

The next logical evolution after the VRU would be to marry telephony services with computer data services. Computer telephony integration (CTI) makes this possible by using a caller-ID interface to capture the caller's telephone information and match this information to a computer's database.

When an agent responds to a call, the computer screen instantly displays all information pertaining to the caller's account. Having this information readily available saves time and money. The customer may also be pleased that customer contact information, sales data, and technical support history are immediately available (Figure 9.3).

CTI can also be integrated into PBX or ACD terminals, as shown in the figure.

Voice Mail

Voice mail is available from a PBX, Centrex, or, for consumers, the central office itself. A voice mail system comprises the following:

- Memory—Stores analog voice digitally
- Speech coders—Convert the analog voice to digital, and vice versa; compression algorithms are generally used to keep file sizes relatively small

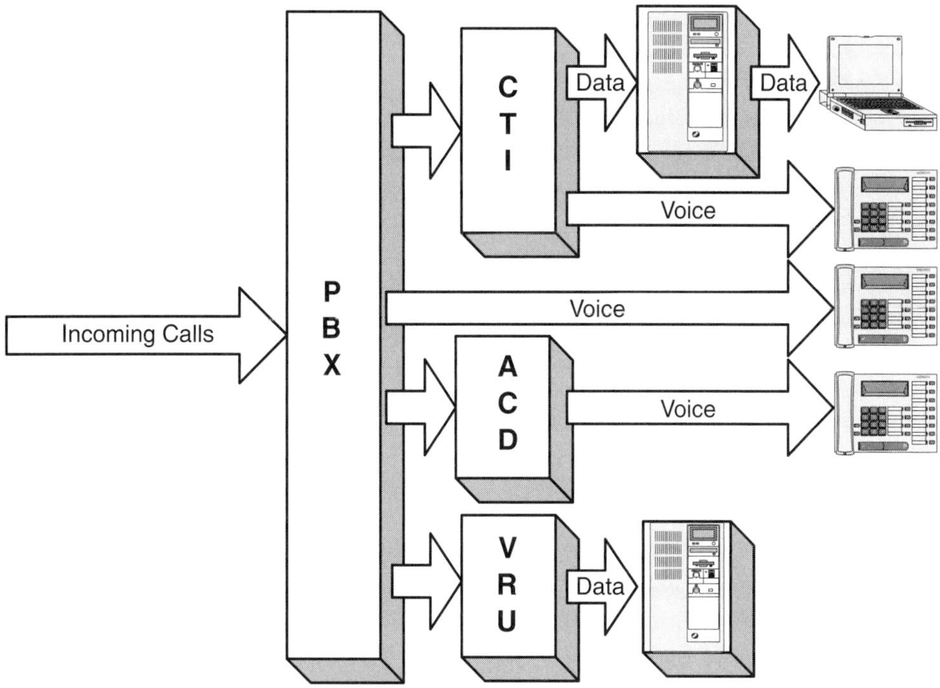

Figure 9.3
PBX with ACD, VRU, and CTI.

- Interface cards—Connect the switching equipment and voice mail speech coders

Voice mail systems generally have a limited memory, so you must delete voice messages at least once a week or the system will delete them for you. An LEC will provide you with voice mail service for a monthly fee. A call waiting lamp can be added to your telephone line. The lamp indicates that there are voice mail messages in your inbox. If you want to maintain voice mail yourself, you can purchase an answering machine.

Answering Machine

Today's digital answering machines can be purchased for less than $50, and they contain a multitude of features and more memory than ever before. Some stand-alone answering machines have up to four voice mailboxes where individuals can have their own voice mailbox. Obtaining several high-quality answering machines for a small business is an economical, cost-effective solution if you cannot justify the higher costs of a key telephone system or a small PBX.

Only digital answering machines should be purchased. Inexpensive answering machines that use cassette tape are not practical for today's consumer or business environments.

Caller-ID

Caller-ID provides a numerical display of the originating caller's number. There are also enhanced caller-ID or caller-ID with name services displaying both the caller's number and name.

For caller-ID to work, both the originator and the receiver must be using digital trunks linked by SS7. Many calling card and large corporations are not SS7 connected, and their caller-ID may not display on the receiver's unit.

SS7 allows the caller's central office to send a calling party number message (CPNM) and a code specifying whether the caller wants her number to be blocked. The caller-ID information and code are transmitted to the receiver's central office where it is translated to the local form of caller-ID and sent on to the caller-ID receiving unit.

Caller-ID information is sent after the first ring-tone and uses the 1200 baud Bell 202 transmission standard for data transmission and information display.

Manufacturers have innovated caller-ID functionality. There are displays for television receivers, wireless displays, and talking caller-ID displays.

Calling Features

The local telephone company or LEC provides custom calling features in its area of service. Not all features are offered by each LEC. In fact, service features can be activated by codes different from the ones described below, so check with your local telephone company first. If the service doesn't activate after entering the code, try entering the "#" key on your keypad after entering the code (Table 9.1).

Table 9.1 Feature Codes

	Enable		Disable	
Feature	**Tone**	**Dial**	**Tone**	**Dial**
Enhanced Call Waiting			*70	1170
Caller ID Blocking	*82	1182	*67	1167
Anonymous Call Rejection	*77	1177	*87	1187
Three-Way Calling	*71			
Call Forwarding	72#		73#	
Call Forward—No Answer	*92	1192	*93	1193
Call Forward—Busy	*90	1190	*91	1191
Return Call	*69	1169	*89	1189
Repeat Dial	*66	1166	*86	1186
Call Trace	*57	1157		

Some of the features, such as repeat dial, three-way calling, and return call, may already be established on your line as a default. Also note that charges may apply for some of these features, so again, you should speak with your local telephone representative before using a particular feature (Table 9.2).

Table 9.2 Feature Description

Feature	Comment
Enhanced Call Waiting	Keeps incoming callers from reaching a busy signal and helps you avoid missing calls.
Caller ID Blocking	Prevents your name and number from being displayed to people when you call.
Anonymous Call Rejection	Rejects calls from incoming callers who block their caller-ID information.
Three-Way Calling	Enables you to talk with two different callers at the same time.
Call Forwarding	Sends incoming calls to the number of your choice when you are away from home.

Table 9.2 Feature Description (Continued)

Feature	Comment
Call Forward—No Answer	Sends unanswered calls to voice mail or another predetermined number.
Call Forward—Busy	When you're on the phone, sends incoming callers to another predetermined number.
Return Call	Redials the last number that called, whether you answered the phone or not.
Repeat Dial	Redials a busy number so you need not wait by the phone and can do other things.
Call Trace	Traces the last call you've received so law enforcement can identify harassing callers.

Distinctive Ringing

We talked about distinctive ringing in Chapter 2. Let's now discuss some possible applications of this service.

Distinctive ringing is one of those low-cost unadvertised features provided by the LEC that can be very beneficial to a home or small business. For example, in my home, my youngest child has a line with two rings and my oldest has a line with three rings. It's great for me, since I don't have to get up to answer the phone when I know the call is for someone else.

Each family member has an individual number that rings on the main line with different ring patterns. The general cost of this service is approximately $3 dollars a month for each additional number. This is a bargain, especially when an individual telephone line can cost between $15 to $20 dollars a month.

In a small business, you may want to use distinctive ringing service to identify a sales call, as opposed to a technical support or vendor call. Each different call type can be identified by a different ring pattern.

Callplex

Callplex is a simple, low-cost, effective device that screens telephone calls and supplements caller-ID. Telemarketing intrusions and unwanted calls are stopped by Call-

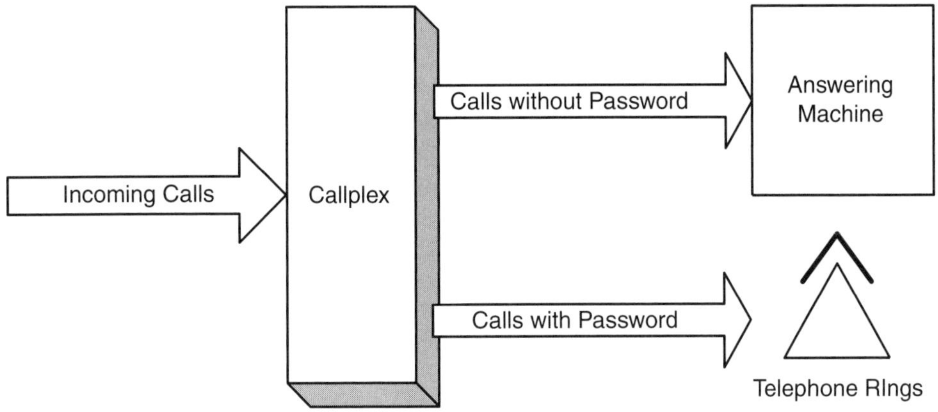

Figure 9.4
Callplex. Courtesy of Callplex (http://www.callplex.com).

plex, but your friends and business associates can get through with the proper pass code. A malicious caller intending to harass or defraud you can't make your phone ring.

In a small business situation, you may want to have a hot-line for special business partners or vendors. When they use the password you give them, the phone rings as usual. Without a password, the call is sent to an answering machine or simply dropped (Figure 9.4).

Long distance callers, who appear as "Out of Area" or "Unavailable" on a caller-ID display, can't be mistaken for unwanted callers if they have your password.

CORDLESS DEVICES

Cordless devices combine the concepts of unguided signal transmission and standard telephone technologies into a powerful, user-friendly device. Cordless telephones provide short-range wireless services without a physical wired connection. Other cordless devices such as pagers can extend your telecommunications needs even farther.

Analog Cordless Telephones

A cordless telephone is actually two separate devices: a base station and a remote unit. A base station requires standard AC power and a 2-wire telephone connection. The base station also has a built-in battery charger to keep the battery in a fully charged state when possible. Both the base station and the remote unit can transmit and receive with each other, using two sets of frequencies or full-duplex transmission.

Always be aware that a cordless telephone is a radio transmitter. Anyone wishing to intercept your analog transmission can always park a car in front of your home or office and attempt to tune in your cordless telephone frequency by using a scanner or receiver. Sometimes, other cordless telephones in your area can pick up the conversation. If you are still unconvinced that privacy is an issue with analog cordless telephones, consider that an intimate telephone conversation between Prince Charles and Lady Camilla Parker Bowles was intercepted by the gossip tabloids in the early 1990s. Amazingly enough, a car outfitted with a scanner and a recording device was all that was needed to pick up the conversation from a location close to the British palace.

Older analog cordless telephones use a single pair of frequencies, and the newer analog phones offer a selection of frequencies designed to reduce signal interference. These telephones operate in the 43–49 MHz range and the new 900 MHz range.

All cordless telephones have a built-in transmission distance limitation. However, with advances in technology and digital phones and spread spectrum technology, frequency ranges have increased, voice clarity has improved, and more importantly, privacy issues have been essentially resolved.

Digital and Spread Spectrum Cordless Telephones

Cordless digital telephones convert the analog voice to a digital signal for transmission. The receiver then converts the digital signal back to analog again. The 900 MHz digital telephones are less susceptible to interference than their analog counterparts. Digital transmission also is much more difficult to pick off than an analog transmission.

The 900 MHz digital spread spectrum telephones are virtually interference proof. These phones offer the highest levels of security with frequencies constantly changing in the entire 900 MHz band.

The 2.4 GHz digital spread spectrum cordless telephones provide the utmost in security, clarity, and freedom from interference. These telephones offer digital encryption in addition to the multiple frequency allocation provided by digital spread spectrum and the reduced interference associated with higher frequencies.

Table 9.3 Paging Devices

Pager Type	Comment	Advantage
Alert only	Pager alerts user with tone only.	Easy to use
Numeric	Pager alerts user with tone and numeric display.	Caller's number displayed
Alphanumeric	Pager alerts user with tone and a numeric/text display.	Caller's number and text displayed
Tone and voice	Pager alerts user with tone and a short (10–20 sec) voice msg.	Caller's voice heard

Public Paging Networks

Receive-only pagers are quite common for both business and consumers alike. The costs for these paging services are minimal as compared with a cellular phone or two-way paging device. Pagers are small and easy to use. A pager can easily fit in a purse or on a belt. Batteries generally last for more than a month, making these devices very economical. Several types of receive-only paging devices are on the market today (Table 9.3).

The coverage for public paging systems can vary significantly. A good paging network will cover a large geographic area. For example, my public pager covers the entire East Coast. If I am traveling to the West Coast, I'll call my paging service and let them know my travel plans. They will simply add my paging codes to West Coast antennas to correspond to my travel plans.

The basic purpose of a receive-only pager is to alert someone that he or she is needed. The paging receiver sets off an alert, which can be visual (flashing light), audible (tone), or vibrating (silent).

When you dial the 800 number of a paging service, you are connected with an automated paging terminal that uses PSTN lines. Encoders convert numbers and or text into codes that can be transmitted. This coded message is sent from the paging service terminal to several antennas, using simulcast transmission techniques. Each antenna has a transmitter that transmits the coded information throughout the paging coverage area. The codes are received by all pagers within a coverage area. However, only the pager with a matching code will process the information and emit a tone or display a message (Figure 9.5).

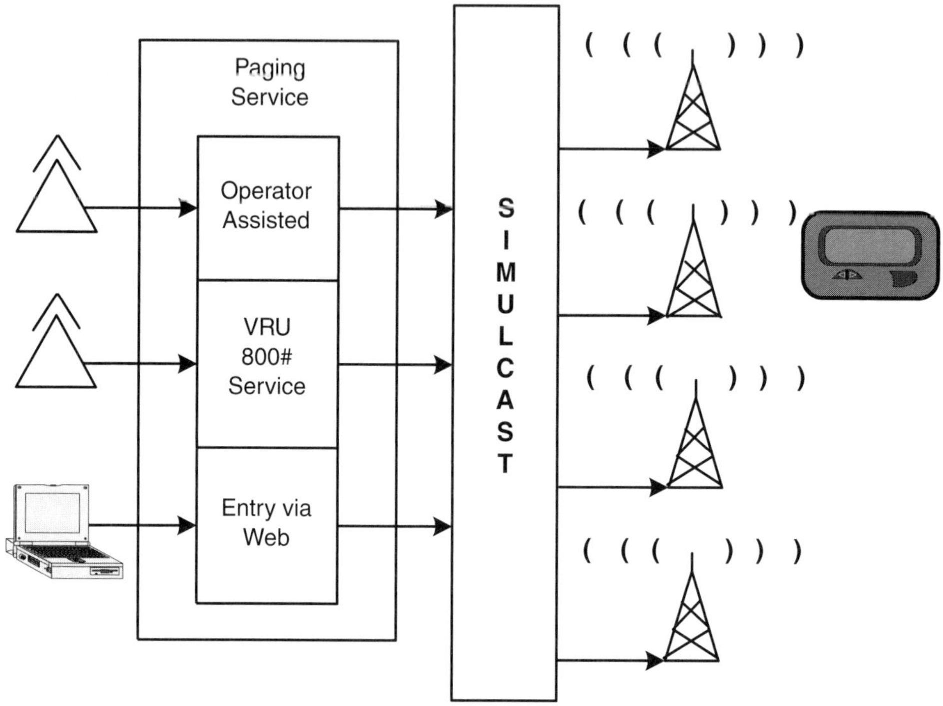

Figure 9.5
Public read-only paging network.

Operator-assisted paging can be used if the recipient has an alphanumeric pager. The operator takes the information verbally and then types in this information for transmission to a pager. Web sites are also available to send an alphanumeric page without an operator.

TIMETABLE OF HISTORIC PAGING EVENTS

1949	Charles Neergard was annoyed by the loud intercom voice paging doctors on the hospital floor and designed a device that would emit a single beep.
1971	The first voice pagers became available.
1980	The first digital numeric display pager becomes available.
1989	The first alphanumeric pagers become available.
1994	Narrowband PCS becomes available, allowing for two-way paging services.

Two-Way Paging Networks

Two-way paging networks use Personal Communications Services (PCS) narrowband spectrum. One channel transmits a reverse-page short message, and another channel receives the page. Two-way paging using the narrowband PCS spectrum sends a reverse page to the antenna from which it received the last transmission.

These pagers have mini-keyboards so that reverse page short messages can be sent immediately.

Private Paging

Private paging devices are used extensively in hospitals, hotels, and restaurants. The coverage is limited to a building or campus that uses a single-site transmitter and antenna. These devices are usually alert-only, for example, an emergency page in a hospital or a restaurant party's page to let the party know that their table is ready.

IMAGING DEVICES

Prices for imaging products have dropped so dramatically in recent years it's worth taking another quick look to see how these devices can enhance existing business.

Many of today's business transactions are done by email and over the Internet. If you want to display your product or service, it is a good idea to have a digital camera available and ready. Once you have stored the images on your hard drive, you can then attach the image to an email message or store the image on a web site.

Scanners have also become an important imaging tool for the office and home environments. To make your office "paperless," you can use a scanner to create an image of a document and thereby provide you with electronic file storage. Scanners can also create an electronic image of a photograph or drawing.

Video projectors have been available for several years. Some recent advancements in projection technology have made video projectors more desirable for business presentations. Today's projectors are lighter and smaller, with an increased light intensity.

Digital Cameras

A digital camera uses no film. Instead, a sensor inside the camera converts the light into electrical charges and stores the information in memory. The sensor itself is called a charge-coupled device (CCD). A CCD contains millions of tiny cells. Each cell can transform light intensity from one small portion of the image into electrons. CCDs can handle low-light functions. CCDs also use a fairly high amount of power, so keep plenty of batteries on hand.

For color, a filter is placed over each cell. This filter breaks up the image into red, green, and blue pixels. Four separate pixels determine the color of a cell. The final color for each cell is determined by an algorithm that averages values from each of the closest pixels.

A camera that has more cells available will have a better image. A higher-quality image requires more memory. As a general rule of thumb you should allow at least 600 Kbits for each mega-pixel of data. For example, a 3 mega-pixel camera should have at least 1,800 Kbits available for image storage.

Almost all digital cameras can use image compression to store these files in less hard-drive space. The JPG and GIF formats use compression; the TIFF format may not.

The maximum image resolution and the number of pixels for the camera have different meanings. For example, a 2.1 mega-pixel camera accounts for some overhead in image processing. Therefore, the actual number of cells used is the maximum image resolution or 1600 x 1200 or 1,920,000.

Each camera manufacturer uses removable media for storage. Floppy disks, smart cards, compact flash, and CD-RWs are sometimes used. A serial cable is usually provided with the camera to download the images directly on to a computer's hard drive.

Three mega-pixel cameras are becoming the standard camera and can be purchased for several hundred dollars. Two mega-pixel cameras are also available for a relatively small amount of money. Both pricing and standards are changing rapidly in the digital camera industry. Professional series cameras from 6 to 10 mega-pixels are also gaining in popularity with the news media and other professional organizations.

Scanners

Three different types of scanners are used for the home or office:

- Flatbed scanner—Allows you to scan images on flat glass. The document is immobile but the scan head moves.

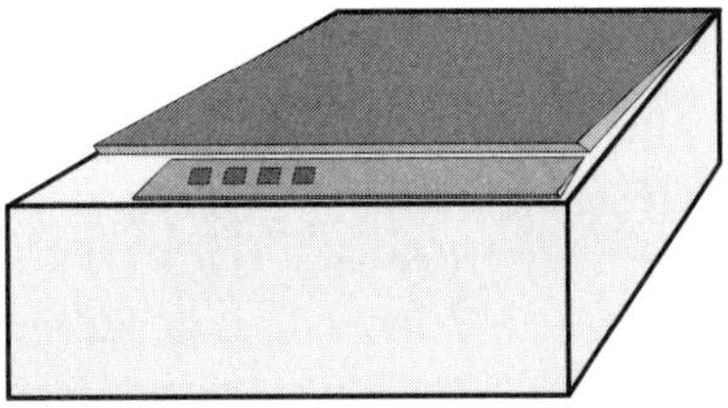

Figure 9.6
Flatbed scanner.

- Sheet-fed scanner—Similar to the flatbed scanner except that the document moves while the scan head remains immobile.
- Handheld scanner—User moves a wand containing a scan head over text.

A xenon lamp or a cold cathode lamp illuminates the image or document. In a flatbed scanner, an entire mechanism moves. This mechanism is called the scan head. The scan head moves across the document by a belt attached to a stepper motor. The scan head contains diode sensors and mirrors with the ability to convert to an electronic format (Figure 9.6).

To obtain color, some scanners use a three-pass scanning arrangement. After the three passes are complete, the scanner software assembles the three images into one image. Whether a scanner passes three times or one time is not critical. The color depth is most important. The color depth refers to the number of colors that a scanner is capable of reproducing. Most scanners sold today can produce 24-bit true color.

The scanner should have a resolution of at least 300 x 300 dots-per-inch (DPI). This number is derived from the number of sensors in the scan head (x-direction) vs. the incremental steps of the stepper motor (y-direction).

Many scanners are advertised with an extremely high DPI. Some are sold with resolutions as high as 9,600 x 9,600 DPI. You should always look for the hardware or optical resolution and not the resolution with software enhancements. Typically, a scanner should have a 600 x 1200 DPI optical resolution.

Scanners require a parallel port, SCSI, or USB interface to obtain access to a computer. SCSI is the fastest but the most difficult to install. A parallel port is the easiest interface to install, but it is also the slowest for scanning.

The cost for an average flatbed scanner is a few hundred dollars. A basic scanner will provide adequate coverage for documents and basic images. For scanning photos, a high-end scanner with the largest hardware DPI is recommended.

Projectors

High-end video projection systems used for presentations and television displays have emerged with a new technology called digital light processing (DLP). DLP sensors contain more than 1 million extremely miniature sized mirrors mounted on support hinges and electrode sensors.

There are two types of mirrors on a DLP sensor. Light shining on the "ON" mirror will reflect through the lens and eventually onto the projection screen. Light shining on the "OFF" mirror will reflect to a light absorber. Each mirror is individually controlled with an electrode tilt system. By using a color filter and by varying the amount of time the DLP mirror is "ON" or "OFF," you create a full color image.

With the mirrors so closely spaced together, it is possible to obtain an extremely sharp image with DLP.

All DLP projectors are extremely lightweight, with the most popular models producing 800 to 1200 lumens of light. This amount of light is more than adequate for conference room or home theater projection. Pricing for this equipment is still relatively high but is expected to drop within the next several years.

ENERGY EFFICIENCY AND SECURITY

With today's technology, there are several relatively low cost solutions to keep your home or office energy efficient and secure. You can cut your energy bill by as much as 25% by purchasing low wattage lighting and energy efficient equipment.

X-10 compatible devices can be a low-cost method of managing your office or home environment. These devices are simple and effective, and they provide the necessary automated switching for lighting and equipment.

Alarms and access control are one of the most overlooked areas of a small office environment. We briefly discuss some of the latest technology for X-10 alarms and proximity access control.

Energy Star

The energy star label indicates that the device is an energy-efficient product. Manufacturers must meet stringent standards to carry this label. PCs with energy star automatically power down to about 50% when a user has stopped using the equipment for a short period of time (Figure 9.7).

Figure 9.7
Energy star symbol.

Lighting technology has changed dramatically during the last 10 years. Replace all incandescent lighting with fluorescent lighting where possible.

Motion sensors are another inexpensive device that can provide savings. These switching devices can minimize your lighting expense by turning unnecessary lights off when no one is in your home or office.

X-10

X-10 technology is an AC power carrier protocol that allows compatible devices throughout the home or office to communicate with each other. This is all done with standard existing wiring. No additional cable or wires are needed to control lighting and equipment. This is the beauty of X-10—it is so simple and easy to use.

Up to 256 codes are available for X-10 addressing. A switching device or transmitter is plugged in to a wall socket. This device has an adjustable code, usually set with a screwdriver. A receiver is also plugged in to a wall socket. The receiver's code must be adjusted to match the transmitter's code. When the switching device or transmitter is set to "on," the code is transmitted over the existing AC power line by a low voltage signal superimposed over the 110 VAC signal. The receiver with a matching code proceeds to switch on the device or lamp connected to it (Figure 9.8).

All outlets in thc home or office receive X-10 signals. Any equipment or devices connected to AC power will receive the low voltage coded information but will discard it as noise.

Manufacturers producing X-10 devices have been very creative in designing their products. Some of these products can be grouped as follows:

- In-wall receivers—Designed to look like a standard dual outlet. One outlet is X-10 switchable and the other is not.

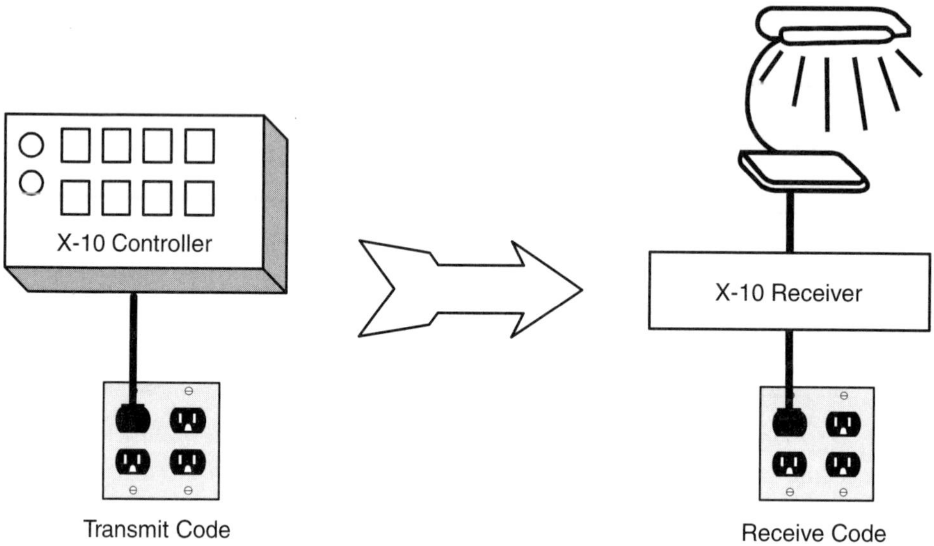

Figure 9.8
X-10 operation.

- Minicontroller transmitters—Contain 16 codes designed to control devices and equipment.
- Dimmer receivers—Dim lamps. These receivers should never be used for devices and equipment.
- Photocell transmitters—Turn lamps on automatically after dusk.
- Timer transmitters—Automatically turn on or turn off lighting and equipment at predetermined times.
- PC transmitters—Have the same functionality as the timer transmitter except that PC software is used to control lighting.

If you are having trouble controlling a receiver from one location and not the other, it may be due to the phase of AC wiring you are using. Most houses are supplied with a 220 volt line that is split into two phases of 110 volts each. A signal bridge will solve the problem by providing a path for the X-10 low voltage signals to cross from one phase to the other.

Sometimes motors or large-current devices can interfere with X-10 transmission. It is best to install an X-10 noise filter on the offending device to prevent unwanted signals from interfering with X-10 signals.

Alarms

You can call an alarm service or an alarm company to meet your needs, but what about alarms using X-10 signals? You can mount a sensor to a window or a door and then connect to an X-10 interface box. Upon a break-in, the "ON" code will be sent to an X-10 alarm receiver, which will then trigger an output device such as a horn, an alarm company signal, or even your pager (with the correct equipment).

Other security devices such as video cameras, intercoms, and wireless sensors can be purchased from home automation or spy shop distributors. Many of these security devices can be interfaced with PCs, so check pricing and see what is available.

Access Control

I don't want to get into a major discussion about locks and keys, but access control for a home office or small business is just as important as access to a computer's server or network.

First of all, keys that cannot be easily duplicated should be used. Keys should be managed. Several manufacturers and locksmiths provide this service. Several companies manufacture locks that have more than one set of keys. If a coworker loses a set, you can easily reset the lock to work with another key without the expense of calling a locksmith.

An electronic proximity switch should also be considered. This system consists of a wall-mounted reader/controller unit and proximity tags that you carry on your keychain. The system assigns a unique code for each tag. The proximity tags are powered from the RF signals transmitted by the proximity reader.

To open a door, the user simply holds a tag in front of the wall-mounted reader. If matching codes are found, an electronic door strike or deadbolt is activated, opening the door.

A single tag can be programmed to open any combination of doors on the proximity switching system. Each tag is unique, so tags cannot be duplicated.

Another similar type of access to the proximity reader is the magnetic stripe reader. A credit card, ATM card, or any other card with a magnetic stripe can be programmed into the system. Up to 100 authorized users (expandable to 1,000) can unlock a door with a single swipe of a predesignated card. No additional keys, pins, or tags are required (Figure 9.9).

Tag Reader Magnetic Card Reader

Figure 9.9
Tag and card readers.

If a tag or magnetic card is lost or stolen, you can remove the ID from the system. Another feature of this type of system is its ability to interface with a PC. Using a PC, you can electronically store entry data in a database for future viewing or reporting.

LOCAL NETWORKING

All networks are created to share information and provide telecommunications between devices or PCs. This time, we look at local networks and servers designed for a small business environment. Let's see why PCs are networked:

- Hardware sharing—Needed so users can share printers, CD-ROM, scanners, etc.
- Program sharing—Needed so only the latest application programs are available on one server
- File sharing—Needed so that users can share their data files on the network
- Communications—Email, Internet, newsgroups, and video conferencing: standard for today's business environment

Peer-to-Peer Networks

Peer-to-peer networking can be configured with bus, token ring, or star topologies. All of the software that you require for this type of networking is included within the operating system. Resources can be shared among individuals.

Peer-to-peer networking is the least expensive local networking and the easiest to install. No PC centralized server is required. Users can operate when other peer computers are down.

This type of networking is for small installations of eight or fewer workstations. Generally, you may want to limit peer-to-peer networking to four users. Each workstation resource may require a password for access of a type called share-level security. Be prepared to keep many passwords saved for future use if you are using this type of access.

Server-Based Networks

Server-based networking enables you to grow your local network, providing a centralized resource that can be backed up and managed. The server itself is typically more powerful than the individual workstations, providing faster response time. The server also verifies individual users and determines their level of access to make the network more secure.

For small business, the cost of server equipment, software, and licensing may be prohibitive, however; you'll need to assess how many employees you presently have and to forecast your potential growth rate.

The Windows NT and 2000 Network operating systems come with a license for five workstations and are available for server-based computing.

Communication Network Servers

A communications server runs specialized software, such as MS Exchange, to communicate. Services such as email and internal group mail are handled by this server. PC workstations access their email from a local network communications server instead of from an email server on a remote network.

Application Network Servers

An applications server can store all of the software needed for business, such as word processing and spreadsheet software. If an application server is installed, each individual workstation does not require its own applications software. This makes updating applications software easier and standardized.

Web Server

A web server enables you to create an intranet web site that is accessible by internal employees. For small business, you may desire to use a web hosting service, which we've already discussed.

For those ambitious entrepreneurs looking to expand their technological horizons, Windows 2000 server comes packaged with Internet Information Server (IIS) 5.0 needed for web development. Be sure that you have a high speed xDSL, cable modem, or dedicated line if you plan on setting up an Internet connection.

TCP/IP and Local Networking

TCP/IP settings on Windows workstations are configured so that the IP address and other settings are received automatically from the DHCP server. If you are using a peer-to-peer network, the DHCP server is typically supplied by your Internet service provider.

If a single connection to the Internet is desired, Windows 2000, XP, or Me will configure themselves automatically, using a feature called automatic private IP addressing (APIPA). If the OS cannot find an IP address from a DHCP server, APIPA will automatically assign one, using the range from 169.254.0.1 to 169.254.255.254.

Windows Domain Network Structure

A domain is all of the user workstations and other network devices that are managed by the network server. This network server or domain controller maintains its own database of user accounts and controls resources such as printers and shared files. A single domain controller can service over 10,000 users.

If more than one domain is in the local network and you need to access another domain, a domain trust is required. The more domains you have, the more trusts you may require. Microsoft has improved the management of domain trusts with its release of Windows 2000 server. A tool called Active Directory enables you to maintain all trusts in a tree-and-branch type of environment. Therefore, with Active Directory, a user in any domain can access any section of the tree, provided that administrator access has been granted.

CONCLUSIONS

This chapter tied up some loose ends. We looked at PBX and other telephone terminating equipment for the first time. We also looked at some equipment intended for the home user, such as caller-ID and voice mail. Depending on your budget and the size of your small business, you may decide to purchase equipment that is expandable. That way, as the technology changes and your business grows, you can still keep your basic equipment setup.

Imaging products are becoming a growth industry, especially for office products. Scanners for paper documents, digital cameras for images, and projectors for slide shows have considerably changed the office environment.

Energy efficiency is just a good idea in general. Turning off equipment and lights when they are not in use saves money. More importantly, energy efficiency is just good for the environment.

Access control, keys, and office security are sometimes overlooked. We examined the possibilities of X-10 devices, proximity sensors, and magnetic card readers for the small-business environment.

Local networking was the last section of the chapter. We now understand the differences between peer-to-peer networks and server-based networks.

This chapter also concludes Part 2. Part 2 covered PC hardware, software, and local equipment. Part 3 now examines new networks and devices. Manufacturers and network designers are looking at new ways of combining existing technologies to improve equipment and networks even further.

Part 3

Telecommunications Technologies for the Future

10 Future Networking Technologies

OK, we've covered everything from telephones to televisions and personal computers. So what's left? Well, the visionaries have been busy with the next wave of telecommunications products and services designed to enhance our living and, they hope, capture market share.

Internet telecommunications technology has been hot and will continue to dominate the interest of both consumers and business professionals alike. Bandwidth is still the major bottleneck for many of us accessing the Internet, so new transmission methods must be devised. These transmission methods must provide us with inexpensive bandwidth that is easy to use and easy to set up.

This chapter is divided into six sections, followed by a salute to the future and my farewell to this book.

Section one looks at potential Internet development starting with Internet2 and the latest version of Internet protocol IPV6.

Section two examines high-speed transmission technologies, which include development of fiber transmission lines and some new backbone cable technology called photonic fiber.

Section three is dedicated to broadcast transmission or enhancements to radio and television as we know it today. Certainly, there has been much discussion regarding HDTV transmission systems, which will eventually replace existing television sets. Also, radio begins to move up to the digital age with digital satellite radio.

Section four looks at wireless broadband technologies that are competing with each other to provide local access to the Internet and video. Local Multipoint Distribution Service (LMDS), Multichannel Multipoint Distribution Service (MMDS), and Wireless Local Loop (WLL) are examined. This section also looks at some high-flying approaches to providing high-speed bandwidth to your home or small business.

Section five focuses on the PC and local access, specifically, Gigabit Ethernet and Bluetooth wireless technology.

The last main section, section six, is the foundation of the future: telecommunications convergence. The dictionary definition of convergence is to move to one point. A good example of this is the public switched telephone network (PSTN), which may one day become part of the Internet with high-speed data networking services.

To the future and beyond...

INTERNET TECHNOLOGY DEVELOPMENT ....

Several problems are associated with today's Internet technology. One, there are too many bottlenecks in the current system. Many users have slow dial-up connections and cannot use streaming multimedia applications. We are also running out of Internet addresses (TCP/IP) so protocol enhancements are needed.

Internet2 capabilities, introduced by then Vice President Albert Gore in 1998, was a first step in experimenting with the next generation Internet. It is a cooperative effort between the federal government, colleges, and universities. Private companies such as Lucent, Cisco, and Nortel are contributors to the Internet2 project.

Today's Internet uses IPV4, which is more than twenty years old. IPV4 has supported us for a long time, but it is starting to show signs of its age. A major problem is the limited number of IPV4 addresses to support the Internet for now and for the future.

Internet2

Since 1996, Internet2 has provided extremely fast connections for colleges and universities only. The purpose of Internet2 is to deploy advanced network applications and technologies, accelerating the creation of tomorrow's Internet. Internet2 is looking to recreate the partnership among educational institutions, industry, and government that fostered today's Internet back in the late 1960s.

The Internet2 network is not available to the general public. The primary goals of Internet2 are to do the following:

- Create a leading-edge network capability for research and development.
- Enable testing of new fiber products and routers.
- Create new network services and applications for the standard Internet.

The network uses high bandwidth backbone fiber similar to the standard Internet. The basic difference between the standard Internet and Internet2 is the number of users. Today's Internet2 network supports 3 million users. In contrast, the standard Internet supports hundreds of millions of users.

The minimum connection speed for Internet2 is 155 Mbps. The network is designed with a minimum number of "hops" between routers, creating even faster connections.

Some experimental testing with Internet2 technology is as follows:

- Medical—3D Brain Mapping: Allows real-time visualization of brain activity during visual and memory tasks. Internet2 allows computers to link together at a high rate of speed to exchange information.
- Medical—Remote Medical Specialist Evaluation: Most medical evaluations can be completed after review of a patient's image by a non-physician at a reading center. Some evaluations, however, require even further review by a supervising specialist. Internet transmission with a guaranteed quality of service and high bandwidth supports real-time interaction among non-physician experts and supervising physicians. Both groups share large image data sets and finalize the evaluation by using Internet2.
- Entertainment—Advanced Digital Video: Internet2 allows for high-quality presentation of video that is not currently available on the standard Internet. Internet2 also provides interactive 2-way digital video, which is not possible with the traditional first-generation Internet services.
- Educational—Remote Instruction and Teaching: Internet2 can link multiple multimedia classroom sessions, providing development of educational materials between two geographically separate institutions. Real-time interaction between students and teachers can be formulated with Internet2.
- Astronomy—Remote Telescope Manipulation: With Internet2, an astronomer in Amsterdam can remotely control a telescope in Hawaii and then participate in an international, full-speed, full-motion videoconference to discuss his or her findings.

Remote interactivity is one of the key features of Internet2. As previously mentioned, doctors can remotely supervise a medical procedure, astronomers can remotely view the contents of a telescope, teachers can remotely instruct a classroom. It is a very powerful scientific and educational medium (Figure 10.1).

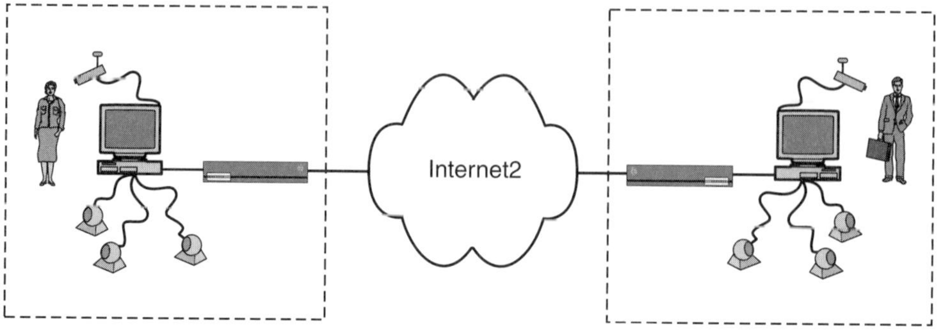

Figure 10.1
Internet2 video conference.

Internet2 may one day become the Internet of the future. In fact, there is already some talk about Internet3 development, with even higher processing speeds than presently available with Internet2.

Internet Protocol Version 6

Internet Protocol Version 6 (IPV6) was developed to fix problems that are associated with today's Internet or IPV4. IPV6 also adds enhancements such as automatic routing and network reconfigurations. This new protocol will eventually replace IPV4, and the two will coexist for several years until a complete transition is made. In fact, IPV6 includes a transition mechanism that provides direct interoperability between IPV4 and IPV6 hosts. It is expected that a full transition to IPV6 will take at least a decade.

The projected use of the Internet in the next few years will be for a much larger group of people. This doesn't include the projected convergence of the telecommunications, entertainment, home appliance, and computer industries, to name a few! (See the section "Convergence" in this chapter for more on convergence).

There are several major differences between IPV4 and IPV6. The first noticeable difference is the address size. IPV6 is 16 bytes long, providing us with a virtually unlimited supply of addresses. A second improvement with IPV6 is the size of the header. The header now contains 7 fields (13 in IPV4). The smaller header size allows for faster packet processing than ever before.

Another improvement with IPV6 is in the area of security. Privacy and security features are always of utmost concern.

IPV6 solves network addressing limitations by replacing IPV4's 32-bit address with a 128-bit address. Despite major advances, IPV6 has been slow to catch on and few commercial products supporting the new protocol are available.

As small pockets of IPV6 networks grow, they will merge into larger and larger pockets. Eventually, all of the IPV6 pockets will become one, and a new Internet will be fully deployed.

BACKBONE TRANSMISSION TECHNOLOGY DEVELOPMENT

A high-speed backbone network is needed for adequate Internet data traffic flow. A weak or overutilized backbone network adds unnecessary bottlenecks and prevents data from getting to its final destination.

Keeping expenses in check is critical since the laying down of fiber, especially in large urban areas, can be costly. New methods of further expanding the capabilities of existing fiber have been produced by the use of increased optics instead of electronics.

Additional scientific experimentation concerning backbone networks includes photonic fiber, which eliminates glass transmission media altogether.

Dense Wavelength Division Multiplexing

Dense wavelength division multiplexing (DWDM) is an optical technique for creating parallel transmission paths on a single fiber. Each path or wavelength can carry as much information as a single-strand optical fiber once carried. So, a 64-wavelength DWDM on one strand of fiber is the equivalent of 64 individual strands of fiber without DWDM.

The concepts regarding DWDM are simple. Remember that prism you needed in high school? Colors are transmitted through the air together and can mix. A prism, however, separates the white light from the sun into a spectrum of colors. In DWDM, each wavelength or color is used as a separate channel (Figure 10.2).

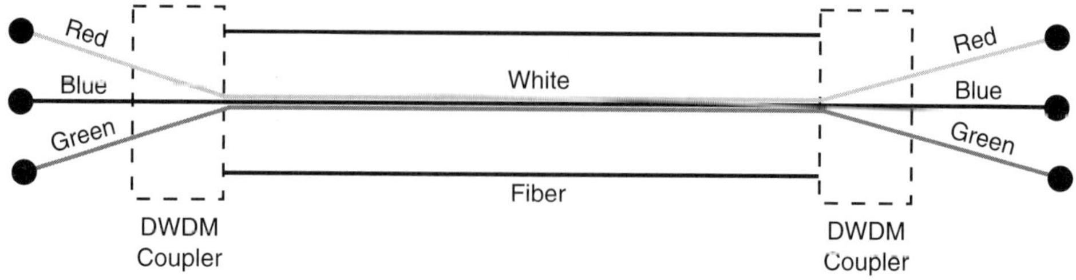

Figure 10.2
DWDM.

As improvements in optical networks and DWDM increase, the cost of backbone fiber bandwidth is expected to decrease, resulting in more optics and fewer electronics in the network.

The costs associated with digging up streets and laying fiber are astronomical. In addition, carriers must negotiate right-of-way, which has become even more expensive in recent years. Generally, it makes economic business sense to use existing resources (fiber) and add DWDM equipment to it.

Ever-increasing bandwidth demands have forced backbone providers to install DWDM equipment. In fact, with DWDM, the bandwidth of fiber can be increased within a few weeks, whereas digging up the streets to install new fiber can take several months or years to accomplish.

The biggest problem with DWDM is the technology itself. Presently, DWDM is in its early stages of development but is rapidly expanding. Each year, more and more wavelengths are made available with newer equipment design. Existing telecommunications carriers must take a gradual approach with DWDM because today's technology will not be tomorrow's state-of-the-art.

Photonic Fiber

For the past 20 years, optical fiber has provided improved telecommunications over longer distances at ever-increasing speeds. There is a physical limit to fiber's transmission speeds. Ideally, we would like to transmit at the speed of light, but this goal cannot be achieved with glass fiber.

The transmission of light through glass fiber varies with its intensity (as opposed to light through free space). More intense light provides a faster speed. If too much light is transmitted with fiber, distortion will result and the signal will degrade.

Scientists are looking at other possibilities for high-speed data transmission besides fiber. Hollow-core photonic fibers or thin air technology may replace fiber altogether. A light is used for signal transmission, and amplification or regeneration is not necessary. Several companies and research institutions are experimenting with this technology, which could someday again revolutionize the Internet.

BROADCAST TECHNOLOGY DEVELOPMENT..........

All of the traditional television and radio services that we have grown up with use analog transmission to provide a signal that can be converted to audio or video. We've already covered the problems inherent with analog signal transmission, so it is just a matter of time before broadcast audio and video signals become digitally transmitted.

There have been some recent attempts to broadcast digital audio via satellite, providing a CD-quality sound to your home or car. High definition television (HDTV) is also making inroads with limited availability. HDTV provides movie-quality video and CD-quality audio right in your living room.

It will be some time before these new broadcast technologies mature enough to replace traditional analog television and radio. Recent attempts by the Federal Communications Commission (FCC) to make HDTV a standard system by the year 2008 are met by much skepticism.

Digital Radio

Satellite radio has been talked about since 1992 when the Federal Communications Commission (FCC) allocated 2.3 GHz for Digital Audio Radio Service (DARS), also known as satellite radio. Very soon, three companies will begin broadcasting music from space via satellites. The three companies are:

- Sirius Satellite Radio—Digital radio broadcast within the United States.
- XM Satellite Radio—Digital radio broadcast within the United States.
- Worldspace—Digital radio broadcast within South America, Asia, and Africa.

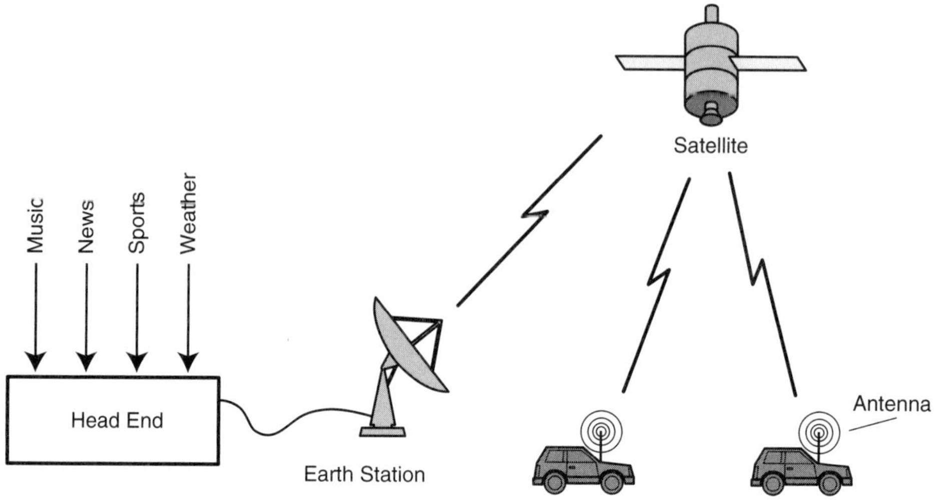

Figure 10.3
Satellite radio.

The concepts for digital satellite radio include an automotive receiver as well as a table receiver. An uplink earth station facility will transmit digital CD-quality music, talk, and news to the satellite 22,300 miles above the earth for reception across the entire United States. Because of the wide footprint of satellite reception, you will be able to drive across the entire country without having to change your radio dial. There will be no static because of the digital signals transmitted. The three companies developing this technology are promising over 100 channels (each) of uninterrupted CD-quality stereo sound with few or no commercials.

Agreements with traditional broadcasters are also being formed to provide services. BBC, CNN, and the Weather Channel are some examples of traditional radio network broadcasting services that will also broadcast via the satellite (Figure 10.3).

For approximately $10 a month, subscribers will decode an encrypted signal that will allow them to receive digital satellite radio. Many car manufacturers have formed agreements with satellite radio providers for installation in their future models. Radio manufacturers including Sony, Sharp, and Pioneer have also begun development on receivers for both auto and home use.

Digital satellite radio has a bright future indeed.

Timetable of Historic High Definition Television Events

1987	The FCC indicates that HDTV standards should be compatible with SDTV service.
1988	The FCC receives 23 different proposals for potential HDTV service.
1990	The FCC requests simultaneous broadcasting of HDTV and SDTV services.
1993	The FCC narrows down the ensuing technology to four contenders and creates a "grand alliance" among all four of them.
1995	The HDTV standard is tested and adopted.

High Definition Television

High definition television (HDTV) has approximately twice the horizontal and vertical resolution of current 525-line analog television broadcast systems. The problem with HDTV isn't the technology itself but its compatibility with existing standards. In 1957, compatibility was a major issue when color television was first introduced. The standard definition television (SDTV) signal needed to be compatible with both black and white and color receivers.

Compatibility between SDTV and HDTV is much more complex than the black and white versus color television issue. First of all, there is the bandwidth; HDTV requires 20 MHz, whereas SDTV requires 6 MHz. Second, SDTV has a 4/3 aspect ratio (almost square), whereas HDTV has a 16/9 wide (letterbox) format. Third, SDTV uses analog signal transmission, and HDTV uses digital signal transmission. Fourth, HDTV lines are scanned sequentially and not interlaced as in SDTV. Because of these issues, HDTV is simultaneously broadcast along with an SDTV signal (on a different set of frequencies).

The design goal of HDTV is to provide more compatibility with computers rather than with the existing SDTV format. HDTV offers 60 frames a second, which is twice the resolution of SDTV. This feature adds to the smooth motion and signal clarity with a movielike quality. Let's also not forget the 5.1 channel AC3 digital sound broadcast included with the video signal.

HDTV is available through digital broadcast satellite (DBS) and from cable TV providers. You must check with either your satellite company or cable provider to determine HDTV compatibility.

Remember our earlier discussions regarding analog and digital transmission? Analog signal transmission is subject to noise and distortion whereas digital signal transmission is relatively noise free and distortion free. For example, an HDTV broadcast originating from Manhattan will look as clear in Jersey City, New Jersey (eastern New Jersey) as in Phillipsburg, New Jersey (western New Jersey). An SDTV broadcast cannot be received well (if at all) in Phillipsburg, New Jersey. Digital signals are either on or off, so unguided transmission noise will not affect the HDTV signal. An extremely weak HDTV digital signal will not work at all.

Original plans, mandated by the FCC, would require SDTV shutdown by the year 2008. Whether this will happen is based largely upon how rapidly HDTV becomes entrenched in our households. HDTV broadcasts are available from cable and satellite companies. As of this writing, HDTV receivers still remain fairly expensive and, when available, are almost always sold with rear projection television systems.

Check out HDTV at your local audio or video electronics store. You don't need any knowledge in telecommunications transmission systems to see the visible difference between SDTV and HDTV.

WIRELESS BROADBAND TECHNOLOGY DEVELOPMENT

We've already talked about our desire to obtain ever-increasing bandwidth for Internet access. Cable modems and xDSL technologies are available today, but they have their limitations. We also looked at Internet satellite applications, which are still expensive, but they do provide high bandwidth services in remote areas that do not have xDSL or cable modem access.

Recall our early discussion of microwave in Chapter 1. Microwave transmission systems were traditionally used for long-haul backbone trunks until fiber arrived on the scene in the early 1980s, virtually replacing them altogether. Microwave technology just couldn't compete with fiber's signal transmission superiority.

Today, there has been resurgence in microwave technology; the emphasis is now with local access. Local Multipoint Distribution Service, Multichannel Multipoint Distribution Service, and Wireless Local Loop services may eventually provide wireless, high bandwidth access to residential and small business users without using a local loop.

Local Multipoint Distribution Service

Local Multipoint Distribution Service (LMDS) is a fixed wireless service capable of transmitting large amounts of data at a very high rate of speed. Microwave radios and antennas are used. One antenna is installed at the client side, and the other antenna is installed at the base or carrier side. The antennas must be line-of-sight directly facing each other and operate in the 25 GHz radio spectrum.

The LMDS base station antenna is generally placed on a high-rise building or tower, and the user's antenna is also placed on a building or tower that provides access.

Because of the cost of equipment and licensing fees, LMDS has not been widely implemented. Another limiting factor of LMDS is its range. Generally, LMDS provides 3 to 5 miles of coverage before the signal begins to degrade to an unusable level.

Multichannel Multipoint Distribution Service

Multichannel Multipoint Distribution Service (MMDS) is also a wireless service supporting up to 99 digital data streams, each transmitting 10 Mbps from a single base station or transmitter. MMDS operates in the 2.5 to 2.7 GHz range, which is substantially lower than the 25 GHz range for LMDS. Equipment for MMDS is therefore less costly than that for LMDS. MMDS can also transmit for more than 30 miles from its base station (depending upon power levels)—significantly more than LMDS service.

The problem with MMDS service is the limited number of channels available. Only 200 MHz is allocated for MMDS use.

MMDS is currently being deployed with some success in large urban areas. These areas already have wired services but because of limited bandwidth copper facilities or high-cost fiber, customers are increasingly turning toward MMDS services to meet their needs.

MMDS and LMDS are both similar in design, but the frequency spectrum allocated for each is different. MMDS will be first in gaining market share with its lower cost and longer range. If MMDS pricing drops and demand for bandwidth remains high, MMDS technology may one day supply movies directly from movie houses to consumers and their home-theater equipment.

Wireless Local Loop

Wireless Local Loop (WLL) is another broadband wireless technology that supplements existing copper-wire services. The service operates in the 24 to 38 GHz range, which means signal distance is limited to a few miles. Like MMDS and LMDS, WLL requires a line-of-sight between antennas (Figure 10.4).

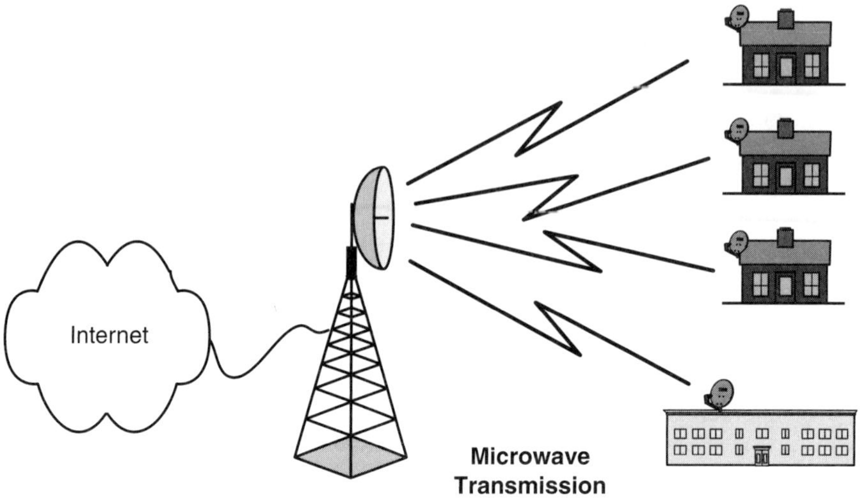

Figure 10.4
LMDS, MMDS, and WLL.

WLL antennas are placed on customer's roofs or small towers that are aimed at a WLL master tower providing service. In some test cases, WLL replaces traditional wiring altogether. For example, a remote community may all share in a T-3 (45 Mbps) that provides wireless connectivity for all of their voice, video, and data needs.

Internet Aircraft

Three companies are presently looking to provide high bandwidth wireless access in hundreds of cities, using a different method of access. An airborne Internet called High Altitude Long Operation (HALO) would use lightweight planes to circle overhead. These planes would act as a relay station (just like a satellite does).

The big difference between Internet satellite and Internet aircraft is delay. A satellite hovering at 22,300 miles above the earth provides a noticeable delay, whereas an aircraft flying at an altitude of 52,000 feet provides minimal delay. The Internet aircraft flying altitude is above commercial aircraft altitudes and inclement weather.

Each city using Internet aircraft could be provided with three planes running 8-hour shifts. In theory, two pilots could split a shift for flying, or the aircraft could run in autopilot mode.

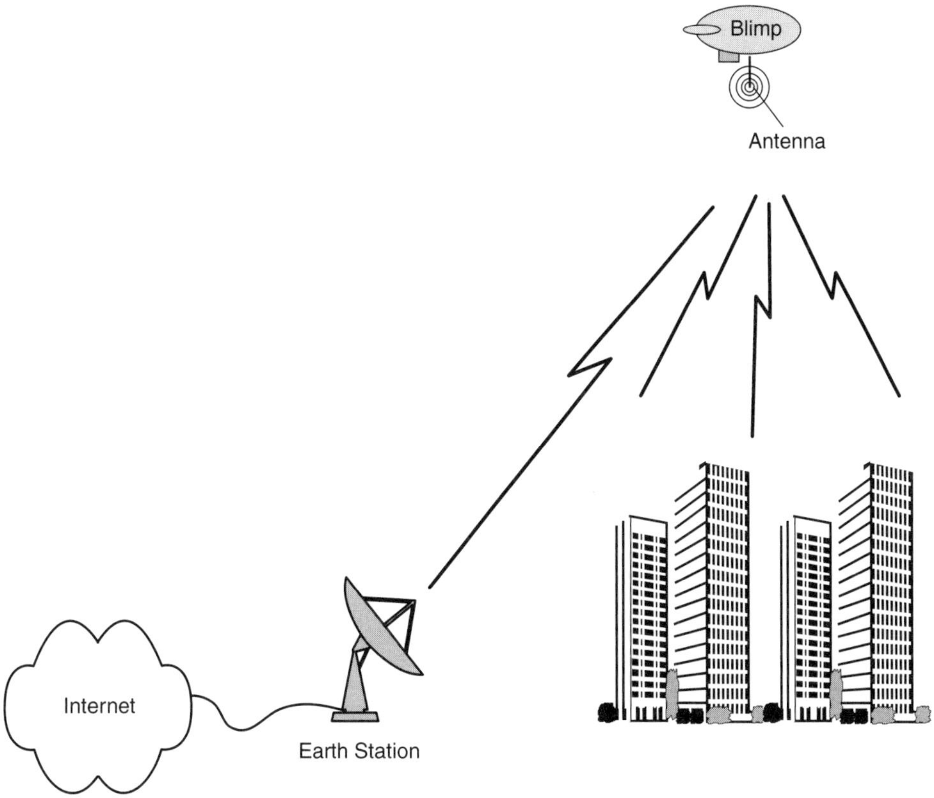

Figure 10.5
Internet aircraft.

An example of such an aircraft is the Proteus project. The National Aeronautics and Space Administration (NASA) has teamed up with Scaled Composites, Inc., to develop such a plane. The Proteus plane is designed as a lightweight, high-altitude, long-duration telecommunications relay platform that can also provide atmospheric-sampling and earth-monitoring missions. The Proteus plane is designed to operate out of general aviation airports with minimal support and low operating costs.

Another example of Internet aircraft is the Internet blimp. A company called Sky Station International plans to fly a blimp powered by solar and fuel cells at an altitude of 13 miles. This unmanned blimp would provide wireless coverage to an area of 7,500 square miles in a geostationary position over a citywide area to provide Internet access (Figure 10.5).

Another even more exotic method of Internet access is the airborne system being developed by AeroVironment and NASA. The Helios aircraft is a solar-powered, unmanned, lightweight plane that can fly for 6 months or more at 60,000 feet without landing. Solar panels provide power, and much of the plane's materials are made out of lightweight Styrofoam and Kevlar. Initial tests during the summer of 1998 and the summer of 2001 have proved to be very successful.

With all of the recent technological advancements using airborne vehicles, the Internet data of the future may be supplied primarily by aircraft or blimp and not by local loops, microwave towers, or cable.

LOCAL NETWORK TECHNOLOGY DEVELOPMENT

We've already examined the potential bandwidth increases for the backbone and wide area networks. Development for local area networks has also been on an upward migration path.

Improvements in Ethernet technology have seen a tenfold increase in speed, and wireless local area technologies are beginning to emerge in the local arena.

Gigabit Ethernet

Both 10 Mbps and 100 Mbps Ethernet have evolved over the last 20 years to become the local networking standard because of their reliability, low cost, and ease of installation. However, streaming multimedia applications are becoming commonplace in the industry, and 100 Mbps may not be adequate in certain situations. Hence, the Gigabit Ethernet standard is born.

The 1 Gigabit Ethernet standard is already used by both corporate and government data networks. This standard does not make existing investments in current local network infrastructure obsolete. Instead, Gigabit Ethernet is designed to serve as the next progressive step for truly high speed local networking. Although Gigabit Ethernet requires fiber optic cable, there are provisions for copper twisted pair as well. Four standards that specify distance and media have been created:

- 1000Base-SX—Standard for short-run, multimode fiber up to 550 m
- 1000Base-LX—Standard for single-mode fiber up to 3 km
- 1000Base-CX—Standard for twin axial copper up to 25 m long

- 1000Base-UTP—Standard for CAT5e cable up to 100 m (uses all four pairs)

If you are looking to upgrade, local switches will require Gigabit Ethernet port connectivity. These new switches operate as a hybrid providing 10 m, 100 m, and 1000 m service.

Gigabit Ethernet provides the same packet framing and the CSMA/CD method as traditional Ethernet systems do. This affords ease of implementation and interoperability with legacy and traditional systems. Minimal training for existing staff and a familiar infrastructure make Gigabit Ethernet a sure bet for an organization deciding to upgrade a local data network.

Before we leave this topic there's just one more thing.

10 Gigabit Ethernet service is also currently in development. 10 Gigabit Ethernet is the next phase of the Ethernet evolutionary cycle. This technology will work with fiber only and will provide the same functionality as 1 Gigabit Ethernet does at 10 times the speed.

Bluetooth

If you take a look at a desktop PC, there are wires connected to a keyboard, monitor, mouse, printer, etc. Other electronic devices such as audio and video equipment have even more cables to connect external devices. Connecting devices with each other sometimes requires complex thinking and special wiring arrangements.

This is where Bluetooth comes in. Bluetooth is named after the king of Denmark (1000 A.D.) who united his country and part of Norway into a single kingdom. This technology is named after him.

Bluetooth is wireless and automatic with the potential capability to simplify electronic connections. Bluetooth will eliminate or reduce local wiring from computers to portable CD-players to audio and video equipment. Bluetooth's radio frequency transmission is at 2.45 GHz and provides a 1 MHz bandwidth for a short distance only.

Bluetooth is designed to transmit by using relatively weak signals for a short range of about 10 meters. Spread Spectrum technology is used. You'll recall from the last chapter that this technology uses frequency hopping techniques to keep security at a maximum with signal interference at a minimum.

It is estimated that the cost of Bluetooth-enabled devices will be an additional $5 to $15 for each device.

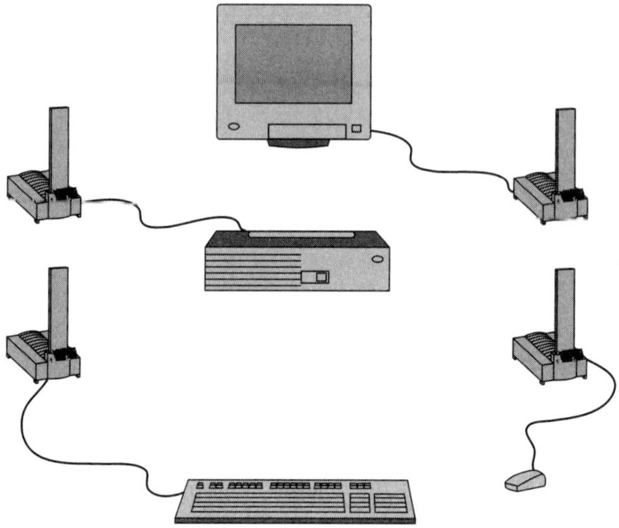

Figure 10.6
Bluetooth-enabled devices.

When any two Bluetooth devices come within range of each other, an electronic signal exchange occurs automatically. The devices attempt to exchange wireless radio frequency data with each other. Obviously, other Bluetooth devices cannot interfere with each other, so a personal area network (PAN) needs to be established. The PAN forms a tiny network. That way, a PC does not interfere with devices from another PC or another electronic device such as audio or video equipment.

All of this wireless connectivity occurs with little or no interactivity from the user. The technology will run at speeds up to 1 Mbps with up to three 64-Kbps voice channels or a master Bluetooth unit broadcasting to seven slave units (Figure 10.6).

The Bluetooth specification has been adopted by over 2000 companies worldwide, with the promise of universal connectivity. Bluetooth and the wireless Ethernet 802.3b standard can coexist, using the same frequency spectrum, according to developers of this new technology.

Only a handful of products with the Bluetooth specification are available. The Bluetooth design is extremely interesting, but there isn't much end-user experience with the technology. Until products are announced and wireless peripherals are tested, there is some uncertainty about interoperability and the features that this technology offers.

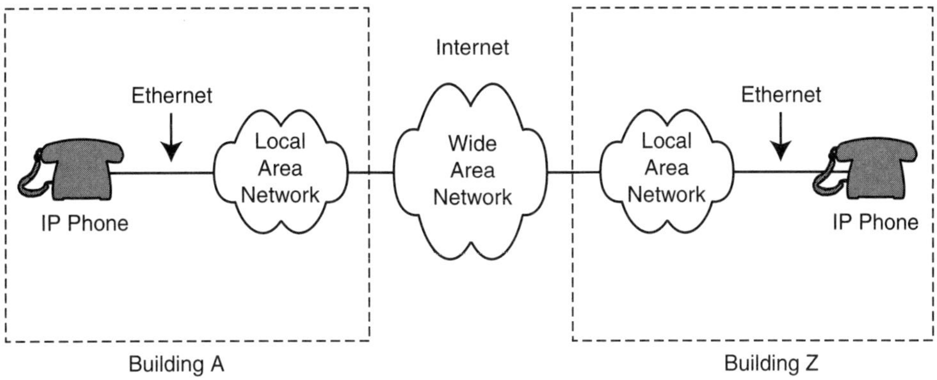

Figure 10.7
VOIP solution.

CONVERGENCE

Now, let's bring together what we know about networks and devices into a single network or single device. This is the idea behind telecommunications convergence. New networks carrying voice, video, and data over IP are beginning to emerge. The transition to a converged network will not be easy and will take many years before a full implementation is completed.

You may ask: Why do we need to converge in the first place? Well, a single network managing all telecommunications needs would be a cost-effective solution as we develop for the future. There are some new considerations like quality of service, lack of features and priority issues, but in general, a converged network using TCP/IP is a good idea.

Both H.323 and Session Initiation Protocol are used today for voice over IP (VOIP) solutions (Figure 10.7). We'll look at the similarities and differences of each of these protocols.

We'll also look at IPV6 and the ramifications of a completely networked environment for the future. Finally, we'll look at some converged devices that are gaining in popularity, specifically, a cellular phone that has audio and video capabilities and a personal digital assistant that functions as a PC and more.

H.323

H.323 is the voice over IP (VOIP) standard that has been recommended by the International Telecommunications Union (ITU). This recommendation provides the technical requirements for voice communications over data networks without any quality of service (QOS) provisions. The standard provides for both point-to-point and multipoint conferencing specifications. Calling features commonly associated with the PSTN (call forwarding and call waiting, for example) are not included with this specification.

The most common use of H.323 is for voice conversation with the Internet as its transmission path. PCs are equipped with microphones, speakers, and the appropriate software to run H.323 with a web browser. The voice quality associated with these calls is usually low, however. The cost for domestic voice services is usually free.

H.323 certainly has its limitations; however, newer VOIP protocols are emerging with even more features and enhancements.

Session Initiation Protocol

Session Initiation Protocol (SIP) is the future of voice telecommunications. Both H.323 and SIP are considered voice over IP (VOIP) solutions. SIP uses a single address for all necessary telecommunications. Computer, telephone, wireless telephone, and personal digital assistant (PDA) can all be used with SIP. SIP can also provide instant messaging, video conferencing, and gaming accessible as a group function.

SIP is a relatively new protocol that determines how voice and data are transported across data networks. It is designed to be integrated with IP services such as video, email, and the Internet, bringing all telecommunications services into one common, integrated platform.

A SIP telephone is sometimes referred to as an Ethernet telephone since the telephone connection is made to an Ethernet jack on a LAN. Each SIP telephone has a TCP/IP address, so you can change telephone device preferences.

Besides network connectivity, a SIP telephone is somewhat different from a traditional telephone. First of all, you must log in to the SIP telephone device much like you would log in for a networked PC desktop device. This login tells the SIP server that you are now available at the telephone you just logged in to. This provides security benefits and allows access to your telephone only to authorized users.

Other features such as call filtering can be used. For example, if you are in a special meeting, you can program your SIP phone to ring only when the call is from your manager. Another new feature is call presence. With call presence you can determine who is logged in on the network and available for voice/data calls. With this information, you can create an audio or video conference among your coworkers.

We already know that SIP services packetize voice for data transmission. If the local network is already heavily congested however, quality of service (QOS) issues may be a major factor. IP networks must be optimized to carry "real time" voice and data together, with minimal delay and errors.

Development of SIP and SIP telephone networks is still in its infancy. Much more study and SIP system implementation is needed before this technology can become a mainstream product. The cost for SIP equipment is still quite high. Cost savings will come from the elimination of PBXs, dedicated lines, and the administrative costs associated with traditional voice networks. Many companies are still reluctant to remove proven PSTN and PBX technologies and replace them with SIP technology. It is anticipated that both technologies will need to coexist for many years to come.

IPV6 and Future Applications

IPV6 is short for "Internet Protocol Version 6," which is a future generation protocol designed to replace the current standard IPV4. Today's IPV4 Internet is almost 20 years old and is having problems. The most serious problem is the growing shortage of IPV4 addresses needed to access the Internet.

IPV6 fixes the address shortage problem and adds many significant improvements to the protocol in areas such as automated network configuration and enhanced routing. With IPV6, some of the IPV4 header information was dropped, creating a simplified packet structure. The 16-byte, 128-bit IPV6 addresses are four times longer than the 4-byte IPV4 addresses. IPV6 is expected to eventually replace IPV4, with the two running together for several years during this transition phase.

One of the most intriguing ideas behind IPV6 is the prospect of assigning IP addresses to virtually everything. Yes. I mean everything! Devices, people, pets, appliances, machines...

People. An extremely small transmitting device can be implanted under your skin. This device would contain your IP address, enabling you to access certain rooms and office buildings without swiping a magnetic card or showing an ID. The IP address could also contain links to information regarding your medical history or financial status.

Pets. An extremely small transmitting device can be implanted under an animal's skin. This device would also contain an IP address for your pet. If the animal is lost, links to information regarding ownership and veterinary history can be obtained.

Appliances. An appliance would contain an IP address with a wired or wireless connection to a local area network in your home. If you are having trouble with the appliance, you would call a service technician who would then diagnose the appliance problem directly over the Internet. The diagnostic would determine whether the entire unit would need replacement or whether certain parts could be fixed and what the labor costs would be. This would all be done without a technician's visit.

Machines. An intelligent soda machine would contain an IP address with a wired or wireless connection to a local area network in an office complex. The intelligent soda machine would be connected to the Internet, and the price of the soda would be determined by answers to the following questions:

- Is the weather hot or cold? In hotter weather, the price of soda would be higher. In colder weather, the price would be lower.
- What is the current inventory of soda in the machine? A fully loaded soda machine would charge less for soda than an empty soda machine.
- When is the next soda delivery date? If soda is to be delivered tomorrow, the price may be lower than if soda is to be delivered next week. An Internet link to the warehouse can determine delivery dates.
- What is the current value of stock for the soda company? A higher value of stock may reflect a higher price for the can of soda. An Internet link to a stock ticker can be used to determine a company's value.
- What are the age and the maintenance requirements for the soda machine? A new soda machine may cost more, raising the price for the can of soda. An older soda machine may require maintenance, which could also be reflected in a higher price for a can of soda.

Personal Digital Assistant

Shrink a laptop computer to the size of a large calculator and you have a personal digital assistant (PDA). PDAs have become quite prevalent in business applications. PDAs use the Windows CE operating system or the Palm OS (Figure 10.8).

Whichever device and operating system is chosen, PDAs have a portability advantage over traditional PCs and laptops. Software exclusively written for PDAs can be similar to software written for PCs. Wireless connectivity can be used to access the Internet or email, and a serial connection can be used to access a traditional PC or laptop.

With devices such as hard drives and memory growing smaller and smaller, it is expected that PDAs will eventually become predominant in the computing industry.

Figure 10.8
Personal digital assistant.

As with any important device as small as a PDA, special security considerations should be taken to safeguard data and passwords.

Mobile Multimedia Phone

The idea of one device for telecommunications voice, data, and video is not new. However, technology has increased so rapidly within the last few years that these concepts are beginning to become a reality.

Many cellular telephones in use today have limited access to the Internet with a micro-browser. Mini-browsers display data such as weather, news, sports, etc. However, today's Internet browsing features are usually very limited.

The widespread acceptance of cellular telephone technology is mainly due to its portability. Voice telecommunications anywhere at any time is truly appealing to both consumers and businesses alike. What about other telecommunications technologies besides voice? Wouldn't it be great to have a converged cellular device that could perform voice video and data services? Let's imagine some of the possibilities:

- Video-conferencing with your coworkers at a bus stop.
- Watching streaming video of a newscast before boarding a commuter train.
- Sending images of a family vacation instantly.
- Receiving MP3 audio of the latest popular music.
- Checking your browser to find the closest fast-food restaurant.

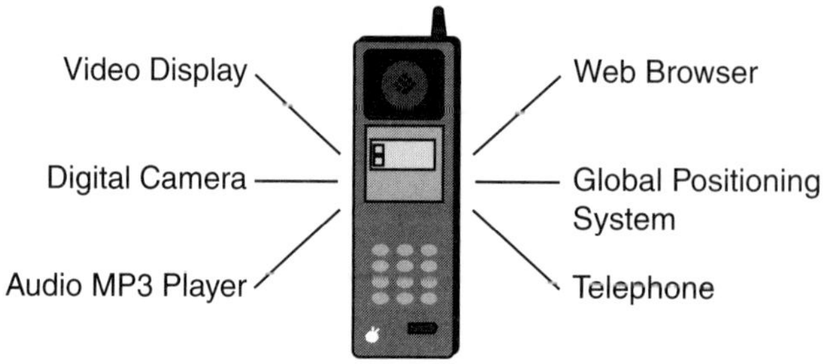

Figure 10.9
Mobile multimedia phone.

The development of a converged cellular multimedia device will not happen all at once. Networks must be developed and retooled, and the appropriate bandwidth must be obtained. Most likely, a gradual conversion from the cellular voice networks of the present to the cellular multimedia networks of the future will happen during the next five years or more.

With a mobile multimedia phone, real-time voice, video, and data will be available to support the desire for any time, any place telecommunications (Figure 10.9).

FINAL CONCLUSIONS

The future of telecommunications technologies has much promise. We can only begin to imagine what changes will be taking place in both local and wide area networks. Surely, the Internet and TCP/IP will play a major role in the development of these networks for the future.

Telecommunications devices keep getting smaller and smaller. As the equipment gets smaller and devices converge into one, we'll see a wireless unit that can provide audio, video, and data in one package. A good example of a converged device is the Dick Tracy two-way wrist TV, which is very close to becoming a reality. WOW!

What else can I say? I'm speechless!

It has been a great technological ride for me. I can't wait to see the devices and networks of the future. I expect there will be plenty more to evaluate and write about during my lifetime!

Appendix

Telecommunications Technologies Scrapbook

This section is written with a personal touch and a lighthearted view of telecommunications technologies. You'll find jokes, a crossword puzzle, and cartoons.

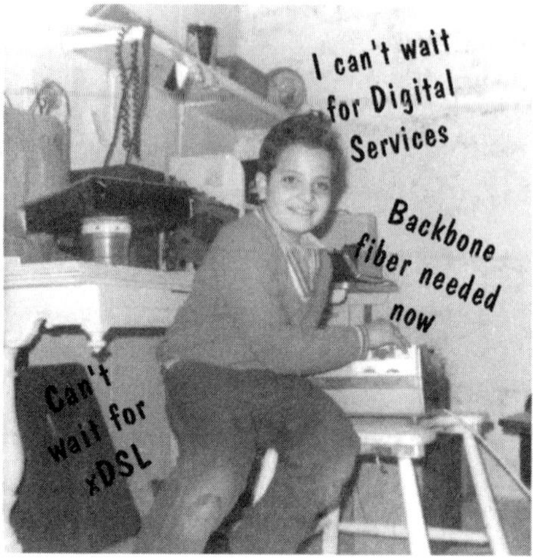

BROOKLYN DREAMS

I can remember dreaming about future technologies as far back as my early elementary school years in Brooklyn, New York. Sometimes, I'd get bored in class and start drawing and making notes in my workbook on possible future technologies. I worked a small job to get enough money for electronic toys and gadgets for my basement workshop. It's always nice to dream.

TELEPHONE MUSIC LESSONS

Music has played such an integral role during my lifetime that I thought why not couple my interest in telephone technology with my interest in music? WOW! Actually, I can provide this type of music lesson at reasonable rates and in fact, we can do it right over the telephone!

- Happy Birthday
  112,163,,,112,196,,,110,8521,,,008,121
  (How old are you now?)
- Jingle Bells
  333,333,39123,,,666,6633,3332232,9,,,333,333,39123,,,
  666,6633,3399631
  (Santa would be proud of you.)
- Mary Had a Little Lamb
  3212333,222,399,,,32123333,22321
  (I can't believe I'm doing this.)

FAMOUS PREDICTIONS

During my elementary and high school education I had some great social studies teachers who influenced me and guided me. I distinctly remember a class where one of my teachers started reading from a book of quotes from famous technological achievers—and whoops!

- 1876. [Western Union]—"This telephone has too many shortcomings to be seriously considered as a means of communication. The device is inherently of no value to us."
- 1899. C. Duell [U.S. Patent Office]—"Everything that can be invented has been invented."
- 1943. T. Watson [International Business Machines Corporation]—"I think there is a world market for maybe 4 or 5 computers."
- 1949. [*Popular Mechanics* Magazine]—"Computers in the future may weigh no more than 1.5 tons."
- 1957. [Prentice Hall]—"I have traveled the length and breadth of this country and talked with the best people, and I can assure you that data processing is a fad that won't last out the year."
- 1977. K. Olson [Digital Equipment Corp.]—"There is no reason anyone would want a computer in their home."
- 1981. W. Gates [Microsoft]— "64K ought to be enough for anybody."
- 1990. J. Dvorak—"I think Windows 3.0 will get a lot of attention. People will check it out and before long they will all drift back to raw DOS."

12 SIGNS THAT YOU ARE IN TROUBLE ....

Sometimes, all of this technology can be overwhelming. Here are some signs that you are in trouble, reprinted with permission of James Qiu (xjqiu@bbcr.uwaterloo.ca):

1. You have never sat through an entire movie without having at least one device on your body beep or buzz.
2. You need to fill out a form that must be typewritten, but you can't because there isn't a typewriter in your house—only computers with laser printers.
3. You think of the gadgets in your office as "friends," but you forget to send your father a birthday card.
4. You disdain people who use low baud rates.
5. You know Bill Gates' email address, but you have to look up your own social security number.
6. You stop saying "phone number" and replace it with "voice number," since we all know the majority of phone lines in any house are plugged into contraptions that talk to other contraptions.
7. You sign Christmas cards by putting :-) next to your signature.
8. You back up your data several times every day.
9. Your wife asks you to pick up some minipads for her at the store and you return with a rest for your mouse.
10. You think jokes about being unable to program a VCR are stupid.
11. You would rather get more dots per inch than miles per gallon.
12. You rotate your screen savers more frequently than your automobile tires.

TELECOMMUNICATIONS TECHNOLOGIES CROSSWORD PUZZLE ....

My wife loves crossword puzzles, so we both decided to work on a telecommunications technologies puzzle together. Here is our final result. Test your knowledge. Good Luck. Answers are on page 326.

Across

1. To transmit signals over a wide area.
6. A resistor or ____ device.
8. Universal Serial Bus.
9. Opposite of bottom.
10. Founder of Microsoft ________.
13. I am (Abbr.) .
14. British Broadcasting Company.
17. Expressed in hertz.
18. Benjamin Franklin discovered electricity, using a _______.
20. A Staten Island telecommunications center.
21. Transmission rate.
23. Analog and ____________ transmission.
25. American Telephone and Telegraph.
27. Very Small Aperture Terminal.
30. Invented the light bulb: ________.
31. Two high-frequency dish antennas pointed directly at each other.
35. _____ Technologies for Small Business.
38. A unit of electrical power.
41. Personal computer.
43. Used in billiards.
44. Alternating current.
45. Lone Flight.
46. The name for video broadcasting and its receiver.
49. A signal with an infinite number of levels per cycle.
50. A device used for voice transmission.
52. A device that is the centerpiece of a star topology.
55. Transmission of text or images.
57. Dual Tone Multifrequency.
60. Asymmetric Digital Subscriber Line.
61. Bert and Ernie's friend.
63. AM and FM _______.
64. Used as a frequency rate in analog transmission.
65. Windows; Office; Net-Meeting; Media Player.
67. AT&T provides voice _______.
69. Music scale.
70. A current of energy.
73. DC voltage can be __________ or negative.
74. Direct current.
75. Inventor of the telephone.

Down

1. A unit of data represented as a 1 or 0.
2. Object-Oriented Modeling and Prototyping (Abbr.).
3. A TV _____.
4. Main web page.
5. Tension between two objects.
6. Wrestling grip.
7. Amplitude Modulation.
11. Short for ampere.
12. A relay antenna orbiting 22,000 miles above the earth.
14. Eight bits equal one ________.
15. DC voltage can be positive or __________.
16. To bow in respect.
19. Southside Johnny and the Asbury _______.
22. Eastern Standard Time.
24. An image or likeness.
25. Medium through which radio signals are sent.
26. To drag.
28. Ready, ________, Go.
29. _________ the Terrible.
32. Small credit-card-sized devices designed for laptops.
33. Disk Operating System.
34. Song-swapping service that uses peer-to-peer computing.
36. Uses CSMA/CD.
37. Thomas Alva Edison.
39. Video cassette recorder.
40. Local ________.
41. 3.1459.
42. Electronic machine that computes data.
47. The basic unit of electrical force or pressure.
48. A wireless or ____________ telephone.
51. Electric Light Orchestra.
53. Point-of-Presence.
54. Its primary protocol is TCP/IP.
56. Turkey's name.
58. A thin strand of glass used for optical telecommunications.
59. A device that converts from chemical to electrical energy.
62. Another name for a CD-ROM drive is an _________ drive.
65. Sounds, pleasing to the ear.
66. Frequency Modulation.
67. Hard drives can use IDE or ______ interfaces.
68. Cathode Ray Tube.
71. Spanish River.
72. Decibel.

SESSION INITIATION PROTOCOL CARTOON..........

Session Initiation Protocol (SIP) may one day provide rules on how telecommunications information is organized, distributed, and transported across networks. I've included several cartoons from Hotsip AB that illustrate this fact (reprinted with permission of Hotsip AB, info@hotsip.com).

In her SIP application Betty notices that her buddy Anton is available for communication. She sends him a message, asking him what"s up.

Anton, being always on the move, has all his communication delivered to his mobile devices. This means his SIP addressis pointing at his phone at the moment. He gets Bettys message, and initiates a voice* session with her.

*In the early days, this was called telephony.

Anton knows the perfect date for Betty. So while switching to his PDA, he can still be reached through his SIP address, now pointing towards this device. Now he can continue his session with Betty. They switch to video and invites Victor, who has his SIP address pointing at his laptop.

Victor is shy, so to brake the ice Betty invites a few friends into the session. Being so many online at the same time, Victor ask the others if they"re up for playing a game. They add the game to their ongoing video session. Now Betty can peek at Victor while slaughtering her other friends in the game.

CLIENT - SERVER

As Betty knows Anton's SIP address, anton@hotsip.com, her SIP client can ask the right SIP server where Anton is. Anton's SIP client has told the SIP server, hotsip.com, that he is online on his mobile. Now Betty's client can send the message to Anton.

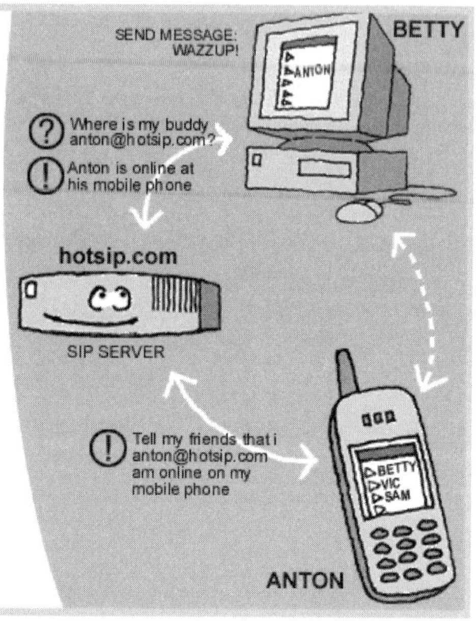

START SESSION

The servers are found by the SIP addres name@server.com. A server, say hotsip.com, is responsible for all the clients that use the addresses ending with @hotsip.com. Clients and servers communicate by asking, answering and reporting to each other. They find each other by the unique SIP address that is supported on the Internet.

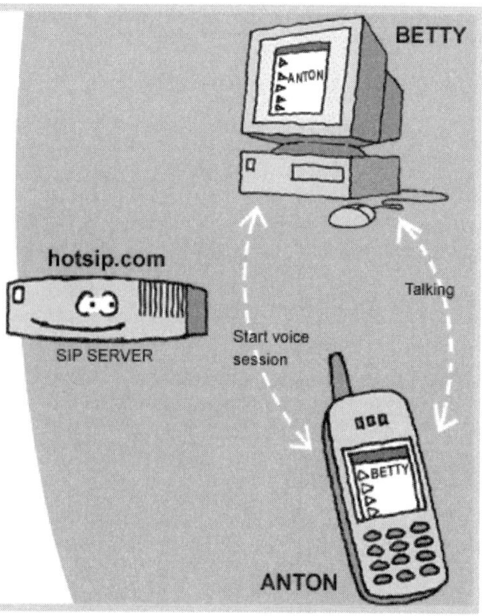

SWITCH SESSION

Anton turns on his SIP client on his PDA. The SIP clients tells the SIP server, hotsip.com, that Anton is now online on his PDA. Then Anton and Betty switches their on going voice session to video. Anton's SIP client asks the tele8 SIP server where Victor is. The server answers that Victor is on his lap top and now Betty and anton can invite Victor to the video session.

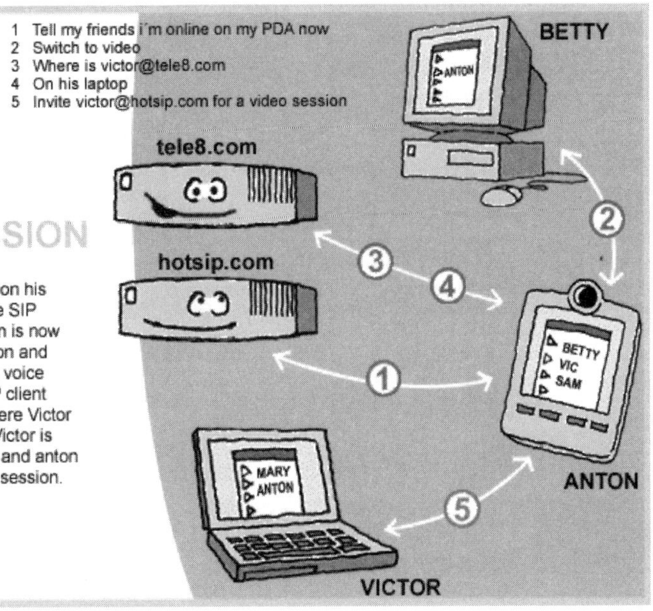

MULTI SESSION

Betty, Anton and Victor adds a game session to their on going video session. They find Madonna and Kenny by asking their servers where they are, as explained before, then invites them to the video and game session. Now five of them can see and hear each other while playing.

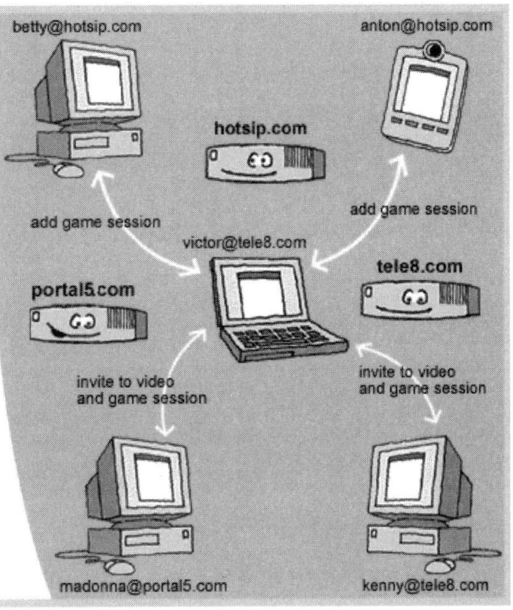

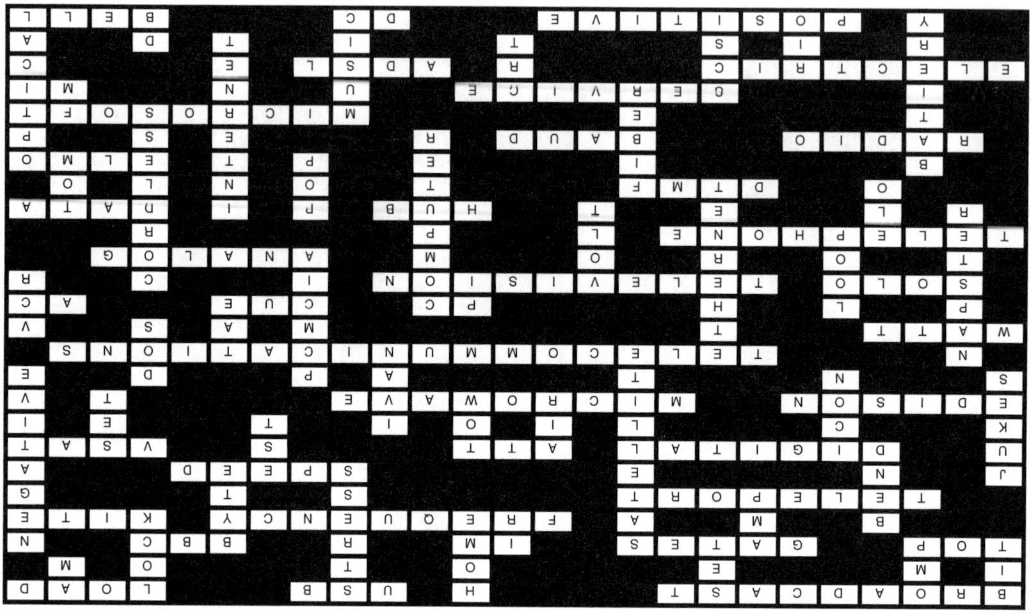

Answers to the crossword puzzle on page 321.

Glossary

100BaseFX

100 Mbps Baseband fast Ethernet specification for using fiber.

100BaseTX

100 Mbps Baseband fast Ethernet specification for using copper.

Accelerated Graphics Port (AGP)

A computer interface used for video monitors, providing a direct link to main memory for video. AGP provides a high rate of speed needed for 3D and high-end graphics.

Alternating Current (AC)

An alternating signal provided by a utility as a power source. This signal always changes from positive to negative.

American Wire Gauge (AWG)

A measurement standard for copper wire. A higher number represents a smaller wire.

Ampere

A measurement of electrical current flowing through a wire.

Amplify

An electronic circuit designed to increase an input level to a desired output level.

Amplitude

Of a sinusoidal wave, the maximum value of a field quantity in space or time. The amplitude of an electrical signal is usually measured in volts.

Amplitude Modulation (AM)

The process by which a carrier wave (frequency) is caused to vary in amplitude by an incoming signal containing information.

Analog

A signal that contains an infinite amount of levels.

Analog-to-Digital Conversion

The process of encoding analog sound waves into a stream of binary digits useful for digital signal transmission.

Answering Machine

A device that can store and play back voice messages.

Antenna

A means for transmitting and receiving electromagnetic signals.

Area Code

A three-digit code used to designate an out-of-area toll center.

American Standard Code for Information Interchange (ASCII)
A seven-bit binary character code that can use an eighth bit for parity checking.

Asymmetric
Using different speeds for outgoing and incoming transmission.

Asymmetric Digital Subscriber Line (ADSL)
A telephone connection that supports up to 8 Mbps downstream and 1 Mbps upstream. ADSL uses a standard local loop telephone wire with a maximum transmission range of 18,000 feet from a central office.

Asynchronous Transmission
A process that uses a start- or stop-bit method of transmission.

Attenuation
Reduction of a transmitted signal as it travels from origination to its destination. Attenuation can also be called a signal loss and is usually measured in decibels.

Audio Frequency
Signal frequencies that can be heard by the human ear.

Automatic Call Distributor (ACD)
A feature of a PBX (Private Branch Exchange) that can automatically route incoming calls to the proper agent.

Backbone
The section of a network that connects local loops, extensions, or LANs. This is the main section of a network providing high-speed connectivity.

Backup
An alternative system that is available with the main system, for use in an emergency, for example, when a main system fails. In telecommunications, a backup circuit may be essential for business-oriented services.

Backhoe Fade
The cut or disconnect of an underground cable.

Bandwidth
The frequency difference between the highest and the lowest available frequencies.

Baseband
A single transmission over guided or unguided media. Baseband is the opposite of broadband (multiple transmissions over the same media).

Basic Rate Interface (BRI)
An Integrated Services Digital Network (ISDN) service containing two bearer or data channels of 64 KBps and one signaling channel of 16 KBps.

Battery
A device that can convert chemical energy to electrical energy.

Battery Backup
A system or device that provides power in the event of a utility outage. A battery backup is sometimes referred to as an uninterruptible power supply, or UPS.

Baud
The actual rate of peak amplitudes. Baud rate may be used in analog transmission.

Binary
A numbering system that uses only two digits (1 and 0).

Bit
A unit of information, such as a 1 or 0, "On" or "Off," +5 volts or 0 volts (contraction of binary digit).

Bit Rate
The number of bits transmitted in a given time interval, for example, bits per second (bps).

Blocked Call
The result of trunks or backbone networks being unavailable in a switched-services circuit. Can be a fast busy signal.

Bonding
The joining of two channels together to provide more bandwidth. For example, two 64-Kbps ISDN channels are bonded together to form one 128-Kbps channel.

Bridge
A device that connects two networks. The bridging device decides whether the data should pass over the bridge or not.

Broadcasting
A signal distribution to more than one station or end terminal.

Browser
A software program that enables a user to view or browse Internet and intranet sites by using point-and-click mouse technology.

Bus
A topology connecting stations in series with each other; usually connected to one main wire.

Busy Hour
The hour of the day when the most telephone calls are made.

Busy Signal
A signal used by the PSTN to indicate that a telephone number a user has dialed is in use.

Byte
Eight bits; also called an octet.

Cable
Guided media that uses copper conductors for transmission.

Cable Modem
A modem used by a cable TV company to provide high-speed Internet access.

Cable Television (CATV)
A network of coaxial cables connected to a head end and designed to provide television services without antennas.

Cable Stripper
A tool used to remove insulation from copper cable.

Call Forwarding
A service whereby incoming calls are automatically forwarded to a different phone number.

Call Setup
The time required to dial a telephone number and make a connection.

Call Waiting
A telephone company feature that allows someone already talking on the phone to receive another call by pressing the switchhook.

Caller-ID
A telephone company feature that sends caller identification information to a caller-ID device after the first ring.

Capacity Planning
A network design plan. Too much capacity wastes bandwidth; too little capacity results in lost calls or data.

Carrierless Amplitude Phase Modulation (CAP)
A process used with xDSL transmission to suppress carrier frequencies at the central office.

Carrier Sense Multiple Access/ Collision Detection (CSMA/CD)
An access method used by Ethernet transmission systems to check that all is clear before transmitting. If two or more PCs are sending, a collision occurs and a retransmission sequence is initiated.

Carrier Signals
A signal or frequency that carries another modulated signal.

Category 3-7
A data transmission specification used for copper wire. Higher numbers provide transmission at higher bit rates.

Cathode Ray Tube (CRT)
An electron-containing tube that illuminates a screen, providing a moving image.

Central Office (CO)
A building used by the telephone company to house local area telephone switching equipment.

Central Processing Unit (CPU)
The "brains" of a PC; performs complex tasks between internal/external devices and memory.

Centrex
A telephone company service that contains a customer office local switch. This equipment is maintained by the local telephone company.

Channel Service Unit (CSU)
A device used as an interface between a digital telecommunications network and local terminal equipment.

Circuit
A path enabling the flow of electrons from one point to another.

Circuit-Switched
Services that provide a temporary path or connection until complete.

Client/Server
An architecture that enables a desktop or client to access data or programs from a LAN.

Coaxial Cable
A shielded copper cable that has one central conductor surrounded by an insulator.

Colocation
An interconnection agreement whereby a competitive company can share space at a central office to access local loops.

Committed Information Rate (CIR)
The rate that a telecommunications carrier will guarantee in a Frame Relay circuit.

Compact Disk-RW
A compact disk format that reads and writes data optically (650 MB).

Competitive Local Exchange Carrier (CLEC)
A company offering business telecommunications services designed to compete with a local carrier.

Compression
A method of processing more information in the same bandwidth.

Computer Telephony Integration (CTI)
A term used to integrate telephones and computer technologies.

Conductor
An element that allows electrical current to flow.

Conduit
Another name for tubing or a pipe inside a building that provides access for cable and fiber runs.

Convergence
The union of disparate elements into one unit. For example, telecommunications devices converge into one unit such as a PDA. Telecommunications networks converge into one network such as TCP/IP.

Current
A unit of electron flow. In Ohm's Law (I = E/R) where E = voltage, I = current, and R = resistance.

Data Communications Equipment (DCE)
A device that acts as an interface to the network. For example, a modem is a DCE.

Data Link Connection Identifier (DLCI)
An identification number given by Frame Relay services to a connection in use.

Data Terminating Equipment (DTE)
End-terminal equipment.

Datagram
A term used in place of the words packet or frame to identify a connectionless environment such as TCP/IP.

Decibel
A measurement showing the increase or decrease of a signal. A ratio of input power vs. output power measured in a logarithmic scale.

Dedicated Circuit
A private line telecommunications circuit that is exclusively used by an individual customer.

Demarcation
An agreed-on connection point where one telecommunication provider ends its service and the other one begins.

Demodulation
The inverse of modulation. Removes a carrier signal from the needed information.

Dense Wavelength Division Multiplexing (DWDM)
A process by which multiple colors of light are transmitted on a single strand of fiber, providing multiple channels of transmission.

Dial Tone
A signal from the central office to tell users they can begin to dial a phone number.

Digital
A signal that has two possible levels: 1 and 0.

Digital Signal 0 (DS-0)
A signal that uses 64 Kbps as a digitized voice line.

Digital Signal 1 (DS-1)
A signal that uses 1.544 Mbps. Also called T-1 rate.

Digital Subscriber Line Access Multiplexer (DSLAM)
The aggregate distribution device usually located at the central office. Used for xDSL services.

Direct Broadcast Satellite (DBS)
A transmission system that provides digital-quality television services. DBS uses a satellite and a receiver instead of an antenna or cable.

Direct Current (DC)
An electrical current that does not change direction from positive to negative.

Discrete Multitone (DMT)
A transmission format used by xDSL to provide 256 subchannels on a local loop.

Disk Drive
A secondary data storage device that stores and retrieves information from disks in a magnetic digital format. Hard disk drives are usually very large (up to 60 GB) while floppy disk drives are usually very small (1.44 MB).

Distinctive Ringing
A telephone company feature that allows telephone sets to ring differently, depending on the number called.

Divestiture
The sale, liquidation, or spinoff of a corporate division or subsidiary. The breakup of AT&T that was associated with the Telecommunications Act of 1984 is an example.

Domain Name
A name that identifies one or more IP addresses.

Dot-Matrix Printer
A device that uses impact technology to create characters on paper.

Dual Tone Multiple Frequency (DTMF)
A pushbutton telephone. Each pushbutton creates two combined frequencies that are used to signal a number.

Digital Versatile Disk (DVD)
A recording medium similar to a CD-R with a much greater capacity (up to 17 GB per side).

E-1
A European standard for digital circuits. Similar to T-1 but has 30 channels instead of T-1's 24. Both use the DS-0 64 Kbps standard.

Earth Station
A satellite transmitting antenna.

Electromagnetic Interference (EMI)
Interference caused by a spurious radio signal or another magnetic field introducing itself into guided or unguided media.

Encryption
A method for ensuring data security. To read an encrypted file, you must have the appropriate password.

Ethernet
A set of specifications defining a telecommunications medium and access method. For example, Carrier Sense Multiple Access with Collision Detection is an access method used primarily in local area networks. Only one transmission is allowed at a time. If more than one station attempts to transmit, a collision occurs and both stations agree on a resend procedure.

Extranet
A company connection that allows certain external users (Internet) access into an internal (intranet) web site.

Federal Communications Commission (FCC)
A unit of the federal government that works in conjunction with local public service commissions. The FCC is tasked with regulating areas of telecommunications for the public.

Fiber Optic
A thin strand of woven glass that has reflective and refractive properties. Used for lightwave telecommunications transmission.

Firewall
Either software or hardware that prevents unauthorized access to an internal network.

Fractional T-1
A subset of T-1 service containing several DS-0's only. For example, a fractional T-1 with 12 DS-0's can be ordered instead of a full T-1 with 24 channels.

Frame Relay
A virtual private networking service used to provide wide area network data coverage. Data rates range from 56 Kbps to 2 Gbps.

Frame Relay Access Device (FRAD)
A device that acts as an interface between a customer's terminal equipment and the Frame Relay network.

Frequency
Measured in cycles per second, or hertz (Hz). One hertz is one cycle per second.

Frequency Division Multiplexing (FDM)
The process of combining several channels of information into one by using frequency division. The combined channels are demultiplexed at the opposite end into their original form.

Frequency Modulation (FM)
Modulation of the frequency of a carrier wave in accordance with an incoming signal.

Full Duplex
The ability to transmit and receive simultaneously.

Gateway
A device that connects two dissimilar networks using different protocols.

Geosynchronous Earth Orbit Satellite (GEO)
A satellite 22,300 miles above the earth moving in synchronous orbit in relation to the earth.

H.323
An ITU standard for sending voice over the Internet.

Half Duplex
Bidirectional telecommunications providing one direction of transmission at a time.

Head End
A centralized location where cable TV signal processing takes place.

Hertz (Hz)
The number of cycles per second in a waveform. One Hz is the equivalent of one cycle per second.

High Bit Rate Digital Subscriber Line (HDSL)
1.544 Mbps bidirectional transmission through local loops with a maximum transmission range of 8,000 feet from a central office.

High Definition Television (HDTV)
A digital transmission format that provides more than twice the number of scan lines as conventional analog television, affording exceptional picture quality.

HyperText Markup Language (HTML)
Text and graphics formatting that is viewable with a web browser.

In-band Signaling
The sending of information, such as dial tone and ringing, over the same circuit or channel as the voice conversation.

Industry Standard Architecture (ISA)

The 16-bit bus standard used by PCs to interconnect internal devices.

Infrared

Light waves above the spectrum of visible light. Requires direct line-of-sight transmission.

Integrated Drive Electronics (IDE)

A PC interface used for both magnetic and optical storage devices.

Integrated Services Digital Network (ISDN)

An early attempt at providing high-speed data services to the PSTN. Two types of services—basic rate interface (BRI) and primary rate interface (PRI)—are available.

International Telecommunications Union (ITU)

An international standards group based in Geneva, Switzerland. Formerly known as the CCITT.

Internet

A wide area network of computers that originated as the ARPANET. The protocol used is TCP/IP. Internet growth has been phenomenal and currently links millions and millions of computers.

Internet Protocol (IP)

Part of TCP/IP that offers connectionless, packetized, internetwork services.

Internet Service Provider (ISP)

A company that offers direct access to the Internet with high-speed connectivity.

Intranet

An internal web-based network.

Intrastate

Internal (within a state), for example, referring to PSTN calls.

ISDN Digital Subscriber Line (IDSL)

64- to 144-Kbps bidirectional symmetrical transmission through local loops with a maximum transmission range of 18,000 feet from a central office.

Jack

A receptacle that accepts a plug, providing connectivity.

Key Telephone System

A device that has some PBX-like features. Key telephone systems are almost always less expensive than a full PBX.

Ku Band

A band of frequencies used for satellite transmission (126 Hz and 18 GHz).

Laser

A device that emits light (photons) when excited by an electric current.

Laser Printer

A printer that uses laser technology to reproduce text and images in a hardcopy format.

Leased Line Circuit

A private line or circuit that is dedicated to an individual customer.

Load

A resistive device contained within a circuit.

Load Coil

A voice-amplifying device used in the local loop.

Local Area Network (LAN)

PCs connected in a local environment with the ability to share and exchange information.

Local Exchange Carrier (LEC)

A telephone company local access provider that houses switching networks and central offices serving customers.

Local Call

A PSTN call which is within a local area with both end terminals using the same central office.

Local Loop

A 2-wire telephone line that runs from the central office to the end-user's premise.

Local Toll Call

A PSTN intrastate call outside the immediate local area at per-minute rate charges.

Long Distance Call

A PSTN interstate call from one state to another state at per-minute rate charges.

Loss

Reduction of a transmitted signal as it travels from its origin to its destination. A signal loss is also called attenuation and is usually measured in decibels.

Low Earth Orbit Satellite (LEO)

A satellite that orbits the earth at an altitude of 500 miles.

Magnetic Storage

A method of storing data by magnetizing sections of media (usually a disk or tape).

Memory

Information stored in read-only format or read-write format.

Mesh

A network design topology that provides multiple access links for the purpose of station backup.

Microwave

Two high frequency radio antennas (and receivers) pointed directly at each other.

Modem

A modulator/demodulator that converts digital information to an analog network line, and vice versa.

Modified Final Judgment

The final court judgment breaking up the Bell System/AT&T into local and long distance telephone services.

Motion Picture Experts Group (MPEG)

A standard for compressed audio and video.

MP3

A technique used for audio compression.

Multimode Fiber

A fiber that allows more light and therefore more reflection and refraction than single-mode fiber because of its larger core. Multimode fiber is less expensive than single-mode fiber and is generally used for short-haul applications.

Multiplexer

A device with the ability to encode and decode several signals into one transmission path.

National Television Standard Committee (NTSC)

The NTSC set the standards for black-and-white and color broadcast television services in the United States.

Network Interface Card (NIC)

An interface card between the PC and the network.

Noise

An unwanted signal introduced into guided or unguided media. Electromagnetic interference is an example of unwanted noise.

Nyquist Theorem

A theory demonstrating that any analog signal can be converted to digital by a digital sample taken at twice the highest analog frequency.

Ohm's Law

A standard electronics mathematical formula (I = E/R) based on the relationships between current, voltage, and resistance, where E = voltage, I = current, and R = resistance.

Oscilloscope

A device that provides a visual view of a waveform or frequencies, showing different voltage levels (vertical) and cycles per second (horizontal).

Out-of Band Signaling

The sending of signals such as dial-tone and ringing by using a separate signaling channel rather than combining the signal with a voice channel.

Packet

A group of bits containing a header (address), payload (data), and ending flag.

Packet Assembler/Dissembler

An X.25 device or software program that breaks up large files into smaller "chunks" for packet transmission.

Pager

A small device that receives wireless simplex messages. Both numeric and alphanumeric pagers are available.

Parallel Data

Multiple transmissions of digital data over a cable of wire.

Parallel Port

A PCs DB25 connector, usually associated with a printer.

Parity

An error checking method whereby the number of 1's in a bitstream are counted, providing an odd or an even result.

Passive Matrix Display

A flat panel display containing rows and columns of liquid crystal elements connected as a grid. These displays are usually associated with laptop computers.

Peer-to-Peer

A local area network that does not use a server or host PC.

Permanent Virtual Circuit (PVC)

A virtual end-to-end connection made between two nodes on a packet network.

Personal Computer (PC)

A stand-alone computer with all the hardware and software an individual needs to complete a task. Originally developed by the International Business Machines (IBM) Corporation in the early 1980s, the PC has evolved into today's high-powered workstation.

Personal Computer Memory Card International Association (PCMCIA)

A standard for small, credit-card-sized devices designed for laptops. These cards are available as modems, hard drives, memory, and LAN interfaces, etc.

Peripheral Component Interconnect (PCI)
A 64-bit bus standard used by PCs to interconnect internal devices.

Personal Video Recorder (PVR)
A device capable of recording or playing back television-quality signals on a television receiver, using a hard disk.

Phase Alteration by Line (PAL)
European television standard.

Phase Modulation (PM)
A method of modulation that makes the carrier signal shift along with a digital input signal. A cosine phase can be the equivalent of "1," and a sine phase can be the equivalent of "0."

Plain Old Telephone Service (POTS)
Standard telephone company services. Also known as the Public Switched Telephone Network (PSTN).

Point-to-Point
A private line or dedicated circuit that has one originating terminal and one end terminal.

Primary Rate Interface (PRI)
An Integrated Services Digital Network (ISDN) service containing 23 bearer or data channels of 64 Kbps and 1 signaling channel of 16 Kbps.

Private Branch Exchange (PBX)
A local telephone switch used at a customer's office, usually containing voice features. A PBX is maintained by the customer.

Projection Television
A method of projecting a television signal broadcast image. Both rear projection and front projection systems exist.

Protocol
In telecommunications, the rules, processes, and procedures used for the transfer of data. Some examples of protocols are Ethernet, Frame Relay, and ISDN.

Public Switched Telephone Network (PSTN)
Standard telephone company services. Also known as Plain Old Telephone Service (POTS).

Quadrature Amplitude Modulation (QAM)
A coding method used in xDSL local loop applications. It represents bits as a combination of phase and amplitude levels.

Queue
In telecommunications, a waiting time.

Random Access Memory (RAM)
Dynamic read-write memory.

Redundancy
In telecommunications, the same as backup.

Regional Bell Operating Company (RBOC)
Local telephone companies. Seven RBOCs were created after divestiture, compared with 22 Bell Telephone RBOCs before divestiture.

Registration Jack 11 (RJ11)
A 2-wire telephone connector.

Registration Jack 14 (RJ14)
A 4-wire telephone connector.

Repeater
A signal regenerator or amplifier. A device used to take a signal that has degraded and make it new again.

Resistance
Opposition of a body or substance to current flow, dissipating energy. In Ohm's Law (R = E/I) where E = voltage, I = current, and R = resistance.

Router
An internetworking device that determines an optimal telecommunications path. Used in wide area networks.

Satellite
A relay station that orbits the earth. Telecommunications satellites send and receive signals from earth stations and antennas located throughout a region.

Scanner
A device that can convert hardcopy photos, text, and images into a digitized format.

Semiconductor
Elements that are neither a conductor nor an insulator. Elements such as silicon or carbon have special properties that allow only a certain amount of current to flow. Transistors and diodes are examples of semiconductors.

Serial Interface
Singular transmission of data over media one bit at a time. Modems generally work with serial interfaces.

Server
A computer that can provide programs and data to other PCs in a local area network.

Set-Top Box
A device containing a receiver and a decoder used for premium channel services.

Shielded Twisted Pair (STP)
A cable composed of two copper wire conductors, shielded for use in noisy environments.

Signal-to-Noise Ratio
The ratio of desired signal vs. an undesired signal or noise. S-to-N is usually expressed in decibels.

Signaling System 7
A method developed in the 1970s by the Bell System/AT&T to provide out-of-band signaling and enhanced caller features for the PSTN.

Simple Mail Transfer Protocol (SMTP)
An electronic mail protocol standard used by the Internet.

Simplex
Telecommunications in one direction only.

Single-Mode Fiber
A fiber with a smaller core than that of multimode fiber. It allows less light and therefore less reflection and refraction. Single-mode fiber is more expensive than multi-mode fiber.

Small Computer System Interface (SCSI)
A PC high-speed parallel interface for both internal and external components. Examples of SCSI devices are hard drives and scanners.

Sound Card
Computer hardware that converts digital data to audible analog signals, and vice versa.

Spread Spectrum
A transmission method that spreads data over several channels. Used when privacy is of utmost concern.

Star Topology
A topology connecting stations together through a centralized hub.

Statistical Time Division Multiplexer (STDM)
Multiplexing technology that dynamically adjusts according to need.

Store and Forward
A procedure that holds data temporarily and forwards it at a later time.

Stranded Wire
A wire that is composed of many smaller wires.

Surge Protector
A device that filters power and offers protection from a voltage surge or spike.

Symmetrical Digital Subscriber Line (SDSL)
A communications line that supports up to 1.544 Mbps downstream and 1.544 Mbps upstream, using a standard local loop telephone wire with a maximum transmission range of 24,000 feet from a central office.

Synchronous Transmission
A mode of data transmission that uses a master clock to synchronize data between the transmitter and the receiver.

Switchhook
A switch on a telephone device used to open a circuit or close a circuit.

T-1
A standard 1.544 Mbps digital circuit. Also known as a DS-1.

Tandem
A switching office that interfaces between a central office and long distance carriers.

Telephone
A relatively simple device used for telecommunications speech.

Television (TV)
A device used to receive audio/video broadcasting services.

Terminal Adapter (TA)
A device that interfaces with the ISDN network, providing the capability to connect to an end-user terminal or telephone.

Time Division Multiplexing (TDM)
The process of combining several channels of information into one, using intervals of time. The combined channels are demultiplexed at the opposite end into their original form.

Token Ring
PCs wired in a ring topology with each device listening to the wire for a single token revolving around the ring. A token may already have data that is intended for another device to receive. A device can transmit when it captures a token and sends data along the wire.

Transistor
A semiconductor device that can amplify or perform fast switching. The transistor replaces vacuum tubes, which are larger, costlier, and require more power.

Transmission Control Protocol (TCP)
Part of TCP/IP that guarantees that data is delivered in the proper sequence without errors.

Tunneling
Encapsulating packetized data to provide additional security.

Twisted Pair
A pair of solid copper wires used for computer and telephone networks. The wires are twisted to reduce noise.

Uninterruptible Power Supply (UPS)
A battery backup system used to supply temporary power when a power utility cannot provide its service.

Universal Resource Locator (URL)
An Internet address, usually consisting of a domain name and a domain identifier.

Universal Serial Bus (USB)
An interface specification for devices that support plug-and-play technology, making devices "hot" swappable and supporting data transfer rates of 12 Mbps.

Unshielded Twisted Pair (UTP)
Twisted pair without a shield or metal wrap. Most voice and data networks use UTP.

V. Standards
International Telecommunications Union (ITU) standards for analog data telecommunications.

Very High Frequency (VHF)
Radio frequencies with a range of 3 KHz to 300 KHz.

Video Cassette Recorder (VCR)
A device capable of recording or playing back television-quality signals on a television receiver. Uses Mylar tape in a cassette housing.

Voice Grade
A 3 KHz bandwidth line.

Voice Mail
An answering machine service usually supplied by a local telephone company.

Voice over Internet Protocol (VOIP)
A method for the transmission of voice over data networks (including the Internet).

Voice Recognition
A technology that enables a computer to recognize and interpret voice.

Voice Response Unit (VRU)
A device that allows a telephone user to automatically interact with a computer to retrieve information.

Voltage
A unit of electronic pressure or force. In Ohm's Law ($E = I \times R$) where E = voltage, I = current, and R = resistance.

Wide Area Network
PCs connected together in an environment that enables them to share and exchange information beyond a local area.

Wireless LAN
A technology providing wireless connectivity in a local environment. Other names for this technology are 802.3b or WI-FI.

World Wide Web (WWW)
Another name for the Internet.

X. Standards
International Telecommunications Union (ITU) standards for digital data telecommunications.

X.25
A packet delivery service specifically designed for short and bursty data situations where error correction and delivery verification are of utmost importance. Used by financial institutions and for stock transfers.

Bibliography

Author: Beyda, William J.
Title: *Data Communications: From Basics to Broadband,* 3rd Edition
ISBN: 0-13-096139-6
Publisher: Prentice Hall, Upper Saddle River, New Jersey (2000).
Subject: Data transmission systems

Author: Bigelow, Stephen J.
Title: *Telephone Repair Illustrated*
ISBN: 0-8306-4034-7
Publisher: McGraw-Hill Professional Publishing, New York (1993).
Subject: Telephone repair

Authors: Ciciora, Walter, James Farmer, and David Large
Title: *Modern Cable Television Technology: Video, Voice and Data Communications*
ISBN: 1-55860-416-2
Publisher: Morgan Kaufman Publishers, San Francisco (1999).
Subject: Cable television

Author: Dodd, Annabel Z.
Title: *The Essential Guide to Telecommunications,* 3rd Edition
ISBN: 0-13-064907-4
Publisher: Prentice Hall PTR, Upper Saddle River, New Jersey (2002).
Subject: Telecommunications

Author: Goralski, Walter
Title: *ADSL and DSL Technologies*
ISBN: 0-07-024679-3
Publisher: McGraw Hill, New York (1998).
Subject: Digital communications, telephone switching systems

Author: Green, James Harry
Title: *The Irwin Handbook of Telecommunications,* 4th Edition
ISBN: 0-07-135554-5
Publisher: McGraw-Hill Professional Publishing, New York (2000).
Subject: Telecommunications systems

Author: Habraken, Joe
Title: *The Absolute Beginner's Guide to Networking,* 3rd Edition
ISBN: 0-7897-2545-2
Publisher: Que, Indianapolis (2001).
Subject: Computer networks

Author: Held, Gilbert
Title: *Next Generation Modems: A Professional Guide to DSL and Cable Modems*
ISBN: 0-471-35981-5
Publisher: John Wiley and Sons, Inc., New York (2000).
Subject: Modems, digital telephone systems

Author: Maxwell, Kim
Title: *Residential Broadband: An Insider's Guide to the Battle for the Last Mile*
ISBN: 0-471-25165-8
Publisher: John Wiley and Sons, Inc., New York (2000).
Subject: Cable television, broadband communications

Author: Mead, Donald C.
Title: *Direct Broadcast Satellite Communications: An MPEG Enabled Service*
ISBN: 0-201-69582-0
Publisher: Prentice Hall, Upper Saddle River, New Jersey (2000).
Subject: MPEG, satellite television

Author: O'Driscoll, Gerard
Title: *The Essential Guide to Home Networking Technologies*
ISBN: 0-13-019846-3
Publisher: Prentice Hall PTR, Upper Saddle River, New Jersey (2001).
Subject: Home networking technologies

Author: Rowe II, Stanford H.
Title: *Telecommunications for Managers,* 5th Edition
ISBN: 0-13-028486-6
Publisher: Prentice Hall, Upper Saddle River, New Jersey (2002).
Subject: Telecommunications management

Author: Simpson, Colin D.
Title: *Industrial Electronics*
ISBN: 0-02-410622-4
Publisher: Prentice Hall, Upper Saddle River, New Jersey (1996).
Subject: Electronics

Author: Stallings, William
Title: *Data and Computer Communications,* 6th Edition
ISBN: 0-13-084370-9
Publisher: Prentice Hall, Upper Saddle River, New Jersey (2000).
Subject: Data transmission systems, computer networks

Author: Tanenbaum, Andrew S.
Title: *Computer Networks,* 3rd Edition
ISBN: 0-13-349945-6
Publisher: Prentice Hall PTR, Upper Saddle River, New Jersey (1996).
Subject: Computer networks

Author: Zacker, Craig
Title: *The Complete Reference: PC Hardware*
ISBN: 0-07-212516-0
Publisher: Osborne/McGraw Hill, Berkeley, California (2001).
Subject: Computer hardware

Index